4/15/66

Concepts of Real Analysis

Concepts of

Real Analysis

Charles A. Hayes, Jr.

Professor of Mathematics and Chairman of the
Department of Mathematics, University of California, Davis

John Wiley & Sons, Inc., New York · London · Sydney

Library of Congress Catalog Card Number: 64—14987
Printed in the United States of America

Preface

This textbook is an outgrowth of a set of notes written for use in one of the courses in a summer institute for college teachers of mathematics held on the Davis campus of the University of California in 1959 under the auspices of the National Science Foundation. The purpose of that course was to improve the participants' understanding of the real number system and the fundamental limit processes associated with it. That purpose has been maintained in the resulting book. It can be used either as a one-semester terminal course or as a foundation for further study in analysis. In the former connection, it could be useful to the prospective teacher who may one day have to teach elementary calculus and needs to understand what lies behind it.

Having taught advanced calculus using a number of fine books, it has been my experience that they either slur over some aspects of the underlying limit theory or present it in rather piecemeal fashion as the work progresses. This is understandable, for the authors want to get on with the theorems of the calculus. However, it seems to me that it is better pedagogically to treat the real number system and simple limit processes separately and thoroughly before complicating matters by introducing derivatives and integrals. The Riemann integral is itself a complex limit that may be better understood after simpler limit concepts have been mastered. Then it should be possible to make rapid progress through the theorems of calculus. Eventually I hope to expand this work to include the theories of differentiation and Riemann integration of functions of several variables and other related topics.

Although it is essentially self-contained and calls on nothing more than a good working knowledge of algebra to be able to read it, a full appreciation of this book requires some mathematical maturity on the part of the reader apart from purely mechanical skills. He should have studied elementary calculus previously, and it would be helpful if he had at least some experience working with abstract ideas and proving things about them. Because this book was written for people who are not already well acquainted with the subject, and not merely as a compilation of known

v

results, I have tried to give intuitive motivation for the formal expression of each concept. After an idea has been expressed in formal language, its properties are developed formally from that point onward.

It has been my experience over the years that skill in proving mathematical statements is the thing students are most slow at acquiring and that mathematics teachers have most difficulty imparting. Only a relatively small number of students ever gain the ability to give a complete and clear proof of a theorem even when they have a good idea of what they should do. The situation reminds me of a chess player who is able to whittle down his opponent's forces but lacks the skill to checkmate him. Consequently, I consider the exposition of the techniques of proof used in analysis to be as much a function of this book as the exposition of the subject itself, and it has been written with that idea in mind. A rather constant level of detail and rigor is maintained throughout, and no significant parts of the theory have been left to the reader to prove. Even the language used in the proofs is rather consistent so as to avoid possible ambiguity and to emphasize the ideas involved. Although a student cannot learn by example alone, nevertheless it is hoped that, by setting him a good and consistent example, he may follow the pattern in his own work. Numerous exercises of varying difficulty are included. The reader should try his hand at solving them in the same general form as the proofs of the theorems in the text. This kind of practice is indispensable to a real understanding of analysis.

Since it is not possible to develop the subject from its beginnings in full detail, there is always a question of where to begin in a text such as this. I chose to start with a partial development of set theory, including only those things that are used later on in the book. It goes somewhat deeper than most elementary treatments, but is it not developed axiomatically. Chapter 2 contains a list of properties sufficient to characterize the real number system. No attempt is made to construct the real numbers; it is assumed that they already exist. Their properties are discussed in varying detail. The purely arithmetic properties are dealt with rather quickly; no effort is made to deduce all the rules of elementary algebra from them, but some exercises on this subject are included. On the other hand, the well-ordering principle of the positive integers, proof by induction, and the completeness axiom are explored at length. The basic properties that are either stated or proved in the first two chapters form the basis for all the subsequent work, and everything from that point on, with minor exceptions, is proved from them.

Chapter 3 is concerned with the definition and properties of finite and infinite sets in general, and sets of real numbers in particular. The Bolzano-Weierstrass theorem is proved therein. The equivalence of the cardinality of the positive integers and the rational numbers and the uncountability of the real numbers are established. The material on cardinality is not used anywhere in the following chapters and can be omitted if desired. Incidentally, it is in this part of the text that there occur the minor exceptions mentioned in the preceding paragraph.

In Chapter 4 a thorough treatment of the theory of convergent real-valued sequences is given. In Chapter 5 motivation is given for adjoining $+\infty$ and $-\infty$ to the real number system, and a thorough study of the theory of limits of sequences in the extended real number system follows.

Chapter 6 treats definition by induction. This is a subject that is often neglected and consequently poorly understood by most students, even though it is intrinsically interesting and important in many branches of mathematics. Although it is here restricted to defining functions on the set of positive integers, the proof of the legitimacy of the process is given in a manner that generalizes easily to general well-ordered sets. In this book the technique of definition by induction is needed to select special subsequences of a given sequence, so there is a practical as well as aesthetic reason for developing the subject here.

Chapter 7 is concerned with the theory of limits of functions of a real variable, continuity, uniform continuity, and uniform convergence of sequences of real functions. In a sense this chapter represents the goal of this book, and so it is rather thoroughly done. In order to emphasize the important relationship between the theorems on limits of sequences and those on limits of a function of a real variable, most of the theorems on the latter are proved in terms of the former. During the summer institute sessions mentioned earlier, many of the participants remarked that they had no real understanding of the technique of proof by contradiction using sequences with specific properties before taking the course, but they did appreciate and understand it afterward. Their difficulty seemed to be due to lack of stress on the relationship between sequential and real functional limits in their own schooling. Alternative proofs not depending on sequential limit theory are left as exercises, with suitable hints on the procedure to follow.

I would like to express thanks and to acknowledge indebtedness to my wife, who typed innumerable versions of the manuscript, and to Professor Bernard Friedman for his many valuable suggestions for improving the manuscript.

C. A. HAYES, JR.

Davis, California
February 1964

Contents

Glossary of Special Symbols

Elements of Set Theory

The theory of sets, which plays an important role in this text, is a highly formal discipline, and it is impossible to do justice to it in a book of this size. Nevertheless, a rather sound understanding of certain set-theoretic ideas is necessary before the sequel can be comprehended. We will proceed by suggesting an intuitive way of regarding sets and use this as the motivation for a partial development of a formal theory. As the development progresses, the intuitive scaffolding will be correspondingly rendered less and less necessary.

The world we live in is made up of many objects. Occasionally we may wish to consider a particular portion of our universe, that is, we may wish to contemplate certain specific objects in the universe and no others. We shall arbitrarily adopt the fiction that the particular objects associated with such a portion are gathered together somehow and placed in an imaginary container. Quite as arbitrarily, we shall call the container a *set*, and we shall refer to the objects inside it as the *members* or *elements* of the set. Clearly, we can imagine the possibility that several different containers capable of holding the same objects might exist. Since everything so far is purely a figment of our imagination and not linked to any other previously formalized discipline, we are free to decide whether or not we wish to allow this possibility. Again, quite arbitrarily, we decide that we do not. Thus we declare that once we are given certain specific objects in the universe, there is exactly one container that will hold them, that is, *there is one and only one set that has these objects, and no others, as its members.* From the point of view we have just adopted, it follows that if we examine two containers and find them to be the same, then even without looking at their contents, we can be certain that they have the same objects within. Also, if we examine the interiors of two such containers, and if we find the contents are identical, then the containers must be the same. The sense of these statements can be expressed as follows: *two sets are equal if and only if they have the same members.*

We now pause to examine the phrase "certain specific objects" that we used rather freely and loosely above. It is necessary to decide on the meaning we wish to attribute to this expression, since, in accordance with the principle adopted in the preceding paragraph, a set is determined whenever we are given "certain specific objects." It is clear that the meaning we give this phrase will delineate exactly the totality of sets that can be defined or, as we sometimes say, *constructed*. We might, for example, insist that objects can be specified only by listing them. However, this would mean that all our sets would be finite and presumably would have relatively few members. This would not be satisfactory, for it would exclude too many interesting situations. We would like to admit the possibility of defining infinite sets and of defining sets whose membership, though finite, might be impossible to list. For example, we should like to be able to define sets whose members are, respectively, all living human beings and all stars in the Milky Way. Even though it is manifestly impossible to list all the objects concerned in either of these cases, we have a strong intuitive feeling that the expressions "all living human beings" and "all stars in the Milky Way" refer to quite specific objects, and we would like to admit the possibility of defining sets with exactly these objects as members. We can imagine all sorts of similar expressions that we would like to use to define sets.

The characteristic feature of the two descriptive statements in the foregoing examples, and others of similar nature, is that they express a property that is or might be possessed by some objects in the universe, and each refers to all objects possessing the property in question.

If we were to do a thorough job of laying the foundations for our theory of sets, we would have to clarify at this point what we mean by "a property" defining a set. We would have to write down formal rules for the formation of statements of such properties. Once adopted, these rules would determine precisely the totality of all sets that could be defined. Unfortunately, such a program would require a lengthy excursion into the realm of formal logic. We have to be content here to say that we will admit as a definition for a set almost any statement that is unambiguously true or false for any given object in the universe, and the members of a set so defined consist of all those objects in the universe for which the statement in question is true. The use of the word "almost" in the preceding sentence is made necessary due to certain well-known paradoxes that arise if all qualifications are removed. In particular, we have to avoid statements that either directly or indirectly define a set in terms of itself. This last relates to the principle that *no set is a member of itself*. It can be shown that violations of this principle lead to logical difficulties.[1]

---

[1] B. Russell's famous paradox, published in 1903, showed the necessity for this restriction. Intuitive justification for this principle can be found in our interpretation of a set as a container, with its elements inside. If a set were a member of itself, then the container representing the set would have to be inside itself.

It is possible, and even highly desirable, to use set theory as a basis for constructing all of mathematics, in which case all objects of mathematical study are sets. However, we can not spare the time to develop everything from set-theoretic beginnings. We find it desirable to assume that certain things have been developed already; and since it is not likely that a beginning student of the subject will be able to imagine how some of these things can possibly be sets, we may regard certain mathematical entities as objects rather than sets. No harm is done thereby.

The terms *set* and *class* will be used synonymously henceforth.[2] Some writers employ the words *collection* and *aggregate* as well. The reason for using different names is, in part, to avoid rather tiresome repetition. There is nothing to prevent us from defining sets whose members are sets. In fact, in accordance with the remarks in the preceding paragraph, it is quite possible to think of all sets as having nothing but sets as members. When we discuss sets whose members are sets, we shall call them *families of sets* or *classes of sets*, mainly to avoid the awkward repetition *set of sets*.

Since we will be working in a mathematical environment, we will often denote sets and their elements by letters. Usually we will use capital letters to denote sets, but when dealing with families of sets we will frequently use script letters to remind ourselves of this fact. Of course, no logical significance is to be attached to the use of any particular letters or kinds of lettering for any purpose.

It is desirable to have a special symbol to express the idea of membership in a set. For this purpose we use $\in$, which will appear in expressions of the form $x \in A$ and may be interpreted and read as *x is a member of A*, *x is an element of A*, *x belongs to A*, or *x is in A*. To express the negation of $x \in A$ we agree to write $x \notin A$ (read *x is not a member of A*, etc.).

This new notation achieves brevity, but the motivation for introducing it goes much deeper than that. We shall soon formalize, that is, write in formal language, certain properties that we want our concept of membership to possess. To be sure that exactly the desired properties and no others shall be incorporated, it is necessary to write them symbolically. In this way we minimize the possibility of ambiguity or misunderstanding that might arise if they were stated in ordinary language. Once the properties of any concept or concepts have been written in symbolic form, we may proceed to derive their consequences by logical inference, making sure that in so doing we use only the formal symbolic statements of these properties and nothing else. Theorems proved in this way will also be symbolic statements that may be translated into statements in ordinary English about the concepts in question.

---

[2] To some authors who develop set theory formally, these words are not synonymous.

Sets may be defined by phrases in English that describe their members. However, in keeping with the foregoing remarks, we are going to introduce a mechanism for defining sets that lends itself well to the symbolism of mathematics. This device is called the *set-builder* or *classifier*. Its general form is shown below:

$$\{x \mid \text{————————————}\},$$

where the blank space is filled in with a statement, usually involving x, of some property that is admissible for the purpose of defining a set. The resulting expression then defines a set, and, in fact, its members are precisely those objects in the universe possessing the property so expressed. In order to determine whether or not any particular object belongs to the set in question, we replace x everywhere it occurs in the statement occupying the blank space by the name of the object under scrutiny and examine the resulting statement. If it is true, the object belongs to the set; otherwise it does not. For example, consider

$$\{x \mid x \text{ is a positive integer and } 1 \leq x \leq 3\}.$$

This is the set of objects x such that x is a positive integer and $1 \leq x \leq 3$. To determine whether or not 1 belongs to it, we replace x everywhere it occurs in the expression to the right of the vertical bar within the braces by 1 and obtain the sentence "1 is a positive integer and $1 \leq 1 \leq 3$." Since this is a true sentence, we infer that 1 belongs to our set. Similarly, replacing x everywhere by $2\frac{1}{2}$, we obtain the sentence "$2\frac{1}{2}$ is a positive integer and $1 \leq 2\frac{1}{2} \leq 3$," which is false since the first part is false, and so we decide that $2\frac{1}{2}$ is not a member of our set. Of course, we can easily see intuitively that 1, 2, and 3 are its only members, and this can also be proved formally.

The fact that a set is properly defined by a perfectly legitimate statement of definition is no guarantee that one can tell whether or not a particular object belongs to the set, or even whether any object belongs to it. For instance, consider

$$\{x \mid x \text{ is a real number and } x \text{ is not algebraic}\}.$$

This is the set of transcendental real numbers. Only relatively few numbers have been definitely identified as transcendental, although it is also known that, in a sense, virtually all real numbers are transcendental. This is not to be construed as meaning that there is anything ambiguous about the criterion for membership in the set of transcendental numbers; it is rather a failure of human mathematicians to be able to check a given number against that criterion.

Even more striking examples of this kind exist. For an arbitrary positive integer $x > 2$, it is not known whether there exist positive integers m, n, and p such tht $m^x + n^x = p^x$. Denoting by A the set

$$\{x \mid x \text{ is a positive integer, } x > 2, \text{ and there exist positive}$$

$$\text{integers } m, n, p \text{ such that } m^x + n^x = p^x\},$$

we do not know if A has any members or not. In order to prove $x \in A$, or else $x \notin A$, it is necessary either to find m, n, and p satisfying the equation $m^x + n^x = p^x$, or else to prove the impossibility of the existence of such m, n, and p. Since there are infinitely many possible choices for m, n, and p it becomes apparent why it might be difficult to establish either $x \in A$ or $x \notin A$. Of course, there are many sets encountered in mathematics that appear to require a check involving infinitely many possibilities to ascertain membership of a particular object in them. However, in some of these cases the logical character of the defining statements happens to be such that one may still be able to give the explicit form of all the members. The situation differs from one case to another.

There are some peculiarities of the classifier that must now be discussed. The role of the variable[3] x appearing next to the left-hand brace inside the classifier is seen to be essentially that of a name for a perfectly general member of the set being defined, and it is introduced simply to describe the conditions that membership imposes on such a typical member. It seems intuitively clear that any other name for a typical member would serve the same purpose equally well. That is, any other variable could presumably be used in place of x, provided it replaces x everywhere x appears in the defining statement. This is true, with certain exceptions to be noted shortly.

Since x can be so replaced without changing the sense of the defining statement, it is sometimes called a *dummy* variable. In expressions of the types:

> for all x, ——————————————— ,

and   there exists x such that ———————————,[4]

where the blank space is filled in with a statement usually involving x, it may be observed that x also plays a dummy role. The expressions "for all x," and "there exists x such that" are called the *universal* and *existential*

---

[3] In the sequel, most of the letters used are variables. Technically one should list in advance all symbols that are to be called variables, but this is neither practical nor necessary in the semiformal presentations we are giving.

[4] An equivalent form "for some x, ——" is frequently used.

quantifiers, respectively. When the variable x is quantified, or appears at the left-hand side of the classifier, it is said to be a *bound* variable. If x appears in an expression and is not bound, it is said to be *free*. The explicit rules concerning replacement of variables fall in the realm of formal logic, and we can not investigate them here. However, we will state the following principle, which we shall use henceforth: *in any given expression, a bound variable may be replaced everywhere it appears by any other variable, provided the new variable is not already in use in the given expression.* An example may serve to indicate why the exception is necessary.

We let z denote a non-zero real number and define the set $\{x \mid x = 2z\}$. If we replace x by 0 we obtain the false statement "$0 = 2z$," and so we conclude that 0 is not a member. Let us see what happens if we replace x by z. We obtain $\{z \mid z = 2z\}$. Clearly 0 is a member of this set, since "$0 = 2 \cdot 0$" is a true sentence. Thus $\{x \mid x = 2z\} \neq \{z \mid z = 2z\}$, since these sets do not have the same membership. According to our assertion above, we should be able to define the same set if we replace x by any other variable except z, which is already in use in the defining expression. We see that in replacing x by z in the expression "$x = 2z$," we made an effective alteration in the nature of the statement.

Before proceeding, we must have an understanding of *equality*. The full story of this concept belongs properly in the domain of logic. However, we can and do adopt the following point of view with regard to it: two mathematical entities are *equal* if and only if every statement that is true for the one is true for the other and vice versa. In particular, this allows the usual substitution of equals for equals, with due regard for the rules on the replacement of variables in statements and formulas and assures the other properties of equality that are commonly accepted and that we shall use freely in the sequel without further justification.[5]

We now wish to express formally the decision we made at the outset, namely, any sets A and B are to be equal if and only if every member of A is a member of B and vice versa. We shall investigate this to see just what it entails. To start with, suppose A and B are equal sets. Their equality alone guarantees that any property possessed by one is possessed by the other; therefore, in particular, the members of A must all belong to B and vice versa. Thus the "only if" part of the foregoing statement is true just because of our agreement on the nature of equality. However, the "if" part says something that is not a consequence of equality or anything else we have introduced. In accordance with the preceding paragraph, two mathematical entities are equal only if all properties possessed by one of them are possessed by the other. However, in case

---

[5] We are referring here to such statements as the following: for all a, b, c, it is true that $a = a$; if $a = b$ then $b = a$; if $a = b$ and $b = c$ then $a = c$.

the entities are sets, to prove them equal we do not have to look at all their properties; we need only look at the particular property of membership. If they have the same members, then they are equal. In effect, this endows the concept of membership with special powers of discrimination with respect to equality. Since we regard the rules of equality as fundamental and already laid down in logic, these special powers of membership will have to be formulated as an *axiom*, that is, a property simply imposed and not subject to proof.

Before writing this axiom symbolically, we shall discuss the form it finally takes. It is generally not possible to consider the members of A one by one and show they belong to B and vice versa. What we are usually forced to do is take a typical or arbitrary member of A, that is, one with no special properties other than its membership in A, and prove it must belong to B, and then take a typical element of B and prove that it must belong to A, in order to establish the equality of A and B. Formal logic decrees that this must be done by using a variable to denote such an arbitrary member of A, say for example x, and then proving logically that $x \in B$ and conversely. When a variable is used in this or similar situations it is not really appropriate to use the term "arbitrary" with respect to it, for this word is not a meaningful modifier for a variable. Nevertheless, we frequently make such statements as "we take an arbitrary element x belonging to $A \ldots$" just to remind ourselves that we are not dealing with a special case. We are now ready to write our axiom and a related definition and theorem.

1.1 Axiom.[6] For any sets A and B, $A = B$ if and only if for all x, $x \in A$ implies $x \in B$, and $x \in B$ implies $x \in A$.

1.2 Definition. If A and B are sets, then $A \subset B$ (read: *A is contained in B*, or *A is included in B*, or *A is a subset of B*) if and only if for all x, $x \in A$ implies $x \in B$. The notation $B \supset A$ (read: *B contains A*, or *B includes A*) is used equivalently with $A \subset B$. The negation of $A \subset B$ is written $A \not\subset B$ or, equivalently, $B \not\supset A$.

In case $A \subset B$ and $B \not\subset A$, we say that A is a *proper* subset of B.

1.3 Theorem. If A and B are sets, then $A = B$ if and only if $A \subset B$ and $B \subset A$.
Proof. This is an immediate consequence of Axiom 1.1 and Definition 1.2.

We now give two examples to illustrate the technique for proving that one set is or is not contained in another, or that two sets are equal.

---

[6] This known as the *axiom of extension*.

We let

$$A = \{x \mid \text{there exists a positive integer } n \text{ such that } x = 4n\}$$
$$B = \{x \mid \text{there exists a positive integer } n \text{ such that } x = 2n\}.$$

Intuitively, it is clear that A is the set of all positive integers divisible by 4 and B is the set of all even positive integers, and therefore $A \subset B$. However, we want to prove this formally.

We should, perhaps, say a few words about the use of the variables x and n in the definitions of A and B. One might become confused, or possibly have the feeling that it is not quite legitimate to use the same name for a typical element of A as for a typical element of B, particularly at a time when we are about to prove that a typical element of A must belong to B. Since x is a bound or dummy variable in each of these situations, it really does not matter what name we use. If desired, we could replace x by y in the definition of B, for example. Usually we do not trouble to make such changes unless the danger of confusion is acute. The same kind of question might be raised in connection with the use of n. For example, one might have the feeling that if $x \in A$ and $x \in B$, then $x = 4n$ and $x = 2n$, whence $4n = 2n$ and $n = 0$, contrary to the requirement that n be a positive integer. But n occurs in both cases in conjunction with the existential quantifier, and so it is a dummy variable. We could again resort to a change from n to m, for example, in the definition of B. Then if $x \in A$ and $x \in B$, we could write $x = 4n$ and $x = 2m$, where m and n denote positive integers, and they are clearly not equal. This would resolve the seeming difficulty we encountered previously. If we wish to use the same variable n in quantified form in two different situations, as we started out doing, then we must keep in mind that it cannot be thought of as having the same value in each of these situations.

To show $A \subset B$, we take an arbitrary element $x \in A$. Membership of x in A assures the existence of a positive integer n such that $x = 4n$. We let $m = 2n$. Then m is a positive integer and $x = 2 \cdot (2n) = 2m$. But the criterion for membership of x in B is that there exist a positive integer such that x is twice its value. Hence $x \in B$. Now we have shown that if $x \in A$, then $x \in B$, and so $A \subset B$.

To show that $B \not\subset A$, it suffices to prove that not all members of B belong to A. This will be established if we can find just one element of B that is not in A. This is the standard way of proving such statements. In the case at hand, we have only to note that $2 \in B$ and $2 \notin A$ to establish that $B \not\subset A$.

As another example, consider

$$A = \{t \mid t = 0 \text{ or } t = x\}, \qquad B = \{t \mid t = x\}.$$

It is clear on replacing t by x that $x \in A$ and $x \in B$. In this case, however, we can not use x to denote an arbitrary element, for this would be equivalent to replacing the bound variable t by another variable, namely x, already in use in the expression under consideration. In situations of this kind we have to use another variable as the name of a typical element. In the case at hand, the variable t is satisfactory. We see that for an arbitrary element $t \in B$, we must have $t = x$. Since $x \in A$, then $t \in A$. We have thus shown that for all t, if $t \in B$ then $t \in A$, and so $B \subset A$. Since $0 \in A$ and $0 \notin B$ unless $x = 0$, we conclude that $A \not\subset B$ unless $x = 0$. This is the most we can say.

We conclude our remarks on this subject by observing that whenever we prove the equality of sets A and B, we are really establishing the logical equivalence of the defining statements of these sets.

It is convenient to admit among our universe of sets one that has no members at all. We call it the *null* or *empty* set, and we denote it by $\varnothing$.

1.4 Definition. $\varnothing = \{x \mid x \neq x\}$.

It is clear that $\varnothing$ has no elements, since if $x \in \varnothing$, then $x \neq x$, contrary to the rules of equality. Thus for all x, $x \notin \varnothing$.

1.5 Theorem. If A is any set, then $\varnothing \subset A$.

Proof. The argument sometimes used to prove this theorem runs as follows: since $\varnothing$ has no elements, then all the elements of $\varnothing$ certainly belong to A, and hence $\varnothing \subset A$. This kind of reasoning may very well seem unsatisfactory, and so we shall pause to put it and similar "vacuously true" arguments on a sound basis.

We consider arbitrary statements P and Q. From these, we can construct a compound statement "P implies Q," or its equivalent "if P, then Q." According to the rules of logic, which we accept here without scrutiny, the truth or falsehood of the compound statement does not depend on the particular form of P and Q but only on the truth or falsehood of P and Q. Specifically, "P implies Q" is true except in the case where P is a true statement and Q is a false one. In particular, if P is a false statement, then "P implies Q" is true whether or not Q is a true statement.

Returning to the problem at hand, in order to prove $\varnothing \subset A$, we must show the truth of the proposition:

(1)   for all x, $x \in \varnothing$ implies $x \in A$.

But, from the remarks following Definition 1.4, "$x \in \varnothing$" is false; hence by the observation just made, it follows that the compound statement

$$\text{“}x \in \varnothing \text{ implies } x \in A\text{”}$$

must be true, whence (1) is true, and so $\varnothing \subset A$.

We note in passing that the null set could have been defined in ways different from that used in Definition 1.4. For example, suppose we define $\varnothing' = \{x \mid 0 = 1\}$. The sentence "$0 = 1$" does not have a variable in it, yet we must still regard such a statement as one into which x is substituted at all appropriate places in order to determine whether or not $x \in \varnothing'$. Since "$0 = 1$" is false, then such substitution leaves it false, and so we infer $x \notin \varnothing'$ for all x. By the argument used in proving Theorem 1.5, we infer that $\varnothing' \subset A$ for each set A. From this we conclude that $\varnothing' \subset \varnothing$; and from Theorem 1.5 itself we see that $\varnothing \subset \varnothing'$, and so $\varnothing = \varnothing'$.

It is possible to define new sets from old ones in various ways. Some of these constructions are of such importance that they have acquired special names and notations by which they are universally known. We shall give some of them now; there are others we will not give, for we have no particular need for them in this book. It is to be noted that some of the operations we shall define, and that we specifically restrict to sets, can be applied to entities other than legitimate sets, but then the results of such operations may not be legitimate sets. The question of just when such operations lead to sets is too technical for consideration here; it is related to the question, mentioned earlier, as to what kind of statements can be used to define sets. In the cases we shall consider, the resulting entities will be sets, and we shall not concern ourselves with the technicalities of the situation.

1.6 **Definition.** If A and B are sets, then

$$A \cup B = \{x \mid x \in A \text{ or } x \in B\};$$

we agree to refer to $A \cup B$ as the *union of A and B*.

It is clear that $A \cup B$ is the set that corresponds to the intuitive idea of combining the sets A and B together to form a single set.

1.7 **Definition.** If A and B are sets, then

$$A \cap B = \{x \mid x \in A \text{ and } x \in B\};$$

we agree to refer to $A \cap B$ as the *intersection of A and B*.

Clearly $A \cap B$ consists of just those objects that are common to A and B.

1.8 **Definition.** If A and B are sets, then

$$A - B = \{x \mid x \in A \text{ and } x \notin B\}.$$

We see that $A - B$ consists of those objects that are in A and not in B.[7]

[7] In some textbooks, especially the older ones, the union of A and B is denoted by $A + B$, and is called the *sum* of A and B; their intersection is denoted by AB or $A \cdot B$, and is called the *product* of A and B. Some authors do not use the notation $A - B$ unless $B \subset A$; others use the *symmetric difference* $A \triangle B$, which coincides with $(A - B) \cup (B - A)$ and reduces to $A - B$ in case $B \subset A$.

The relation of inclusion and the operations of union and intersection just introduced satisfy conditions resembling in some ways those satisfied by ordinary inequality, addition, and multiplication, respectively.[8] Thus we are led to a kind of algebra of sets that enables us to calculate with sets in a manner similar to the way we calculate with numbers with, however, certain noteworthy differences, as we shall see. The algebra of sets is known as *Boolean* algebra. We now prove some of its basic properties.

1.9 Theorem. If A and B are sets, then $A \cup B = B \cup A$ and $A \cap B = B \cap A$.

Proof. Consider an arbitrary element $x \in A \cup B$. By Definition 1.6, $x \in A$ or $x \in B$; thus $x \in B$ or $x \in A$,[9] and so again by Definition 1.6, $x \in B \cup A$. Hence $A \cup B \subset B \cup A$. Interchanging the roles of A and B in this argument, we obtain $B \cup A \subset A \cup B$, and so $A \cup B = B \cup A$.

Next, consider an arbitrary element $x \in A \cap B$. By Definition 1.7, $x \in A$ and $x \in B$; hence $x \in B$ and $x \in A$, and so by Definition 1.7, $x \in B \cap A$. Thus $A \cap B \subset B \cap A$. Interchanging the roles of A and B in this argument, we obtain $B \cap A \subset A \cap B$, whence $A \cap B = B \cap A$.

The results just obtained are called the *commutative laws of unions and intersections*, respectively.

1.10 Theorem. If A and B are sets, then $A \subset A \cup B$ and $A \cap B \subset A$.
Proof. Take an arbitrary element $x \in A$. From the truth of "$x \in A$," we may infer the truth of "$x \in A$ or $x \in B$;" thus $x \in A$ implies $x \in A \cup B$, and so by Definition 1.6, x belongs to the set $A \cup B$. Hence $A \subset A \cup B$.

Next, take an arbitrary element $x \in A \cap B$. From the truth of "$x \in A$ and $x \in B$" we may infer the truth of "$x \in A$;" hence $x \in A \cap B$ implies $x \in A$. Thus $A \cap B \subset A$.

1.11 Theorem. If A, B, and C are sets, $A \subset B$, and $B \subset C$, then $A \subset C$.
Proof. Take an arbitrary element $x \in A$. Since $A \subset B$, then $x \in B$. Since $B \subset C$, then $x \in C$. Thus we see that if $x \in A$, then $x \in C$, and so $A \subset C$.

1.12 Theorem. If A, B, and C are sets, and if $A \subset C$ and $B \subset C$, then $A \cup B \subset C$; if $C \subset A$ and $C \subset B$, then $C \subset A \cap B$.
Proof. Assuming first that $A \subset C$ and $B \subset C$, take an arbitrary element $x \in A \cup B$; then $x \in A$ or $x \in B$. If $x \in A$, then since $A \subset C$, it follows that $x \in C$; if $x \in B$, then since $B \subset C$, it follows that $x \in C$. Hence in any case, if $x \in A \cup B$, then $x \in C$, so that $A \cup B \subset C$.

[8] This was the reason for the use of $+$ and $\cdot$ to denote unions and intersections, respectively, and for calling them sums and products.
[9] Here and elsewhere, when we deal with connectives such as "or" and "and" we invoke and utilize properties conventionally bestowed on them by logicians. In particular, if P and Q are statements, then the compound statement "P or Q" is true in case P is true, or Q is true, or both are true.

Next, assuming that $C \subset A$ and $C \subset B$, take an arbitrary element $x \in C$. Since $C \subset A$ then $x \in A$; since also $C \subset B$, then $x \in B$, whence $x \in A \cap B$. Consequently $C \subset A \cap B$.

1.13   Corollary.   If A is a set, then $A = A \cup A = A \cap A$.
Proof.   This follows immediately from Theorems 1.10 and 1.12.

1.14   Theorem.   If A, B, C, and D are sets, $A \subset C$, and $B \subset D$, then $A \cup B \subset C \cup D$ and $A \cap B \subset C \cap D$.
Proof.   Applying Theorems 1.10, 1.11, and 1.9 we infer that $A \subset C \cup D$, $B \subset C \cup D$, $A \cap B \subset C$, and $A \cap B \subset D$. Application of Theorem 1.12 completes the proof.

1.15   Corollary.   If A is a set, then $A \cup \varnothing = A$ and $A \cap \varnothing = \varnothing$.
Proof.   By Theorems 1.5, 1.10, 1.14, and Corollary 1.13 $A \subset A \cup \varnothing$, $A \cup \varnothing \subset A \cup A = A$, $\varnothing \subset A \cap \varnothing$, and $A \cap \varnothing \subset \varnothing$. These relations yield the desired conclusions.

The definitions for unions and intersections were formulated for two sets at a time. We now consider three sets, form the union or intersection of two of them, and then unite or intersect, respectively, the result of this first operation with the remaining set. The following theorem tells us something about the independence of the end result with respect to the order in which the operations are performed. In conformity with our expressed aim, it has to be proved using only formal properties of the membership concept. Incidentally, the results of this theorem are known as the *associative laws of unions and intersections*. They bear an obvious relation to the associative laws of numerical addition and multiplication, respectively.

1.16   Theorem.   If A, B, and C are sets, then

$$\text{(i) } A \cup (B \cup C) = (A \cup B) \cup C \text{ and}$$
$$\text{(ii) } A \cap (B \cap C) = (A \cap B) \cap C.$$

Proof.   (i) From Theorems 1.10 and 1.9 we infer that

$$A \subset A \cup (B \cup C), B \subset B \cup C, C \subset B \cup C,$$

whence, applying the same theorems to these relations, we obtain

(1) $A \subset A \cup (B \cup C)$, $B \subset A \cup (B \cup C)$, $C \subset A \cup (B \cup C)$.

From Theorem 1.12 and the relations (1) we see that $A \cup B \subset A \cup (B \cup C)$; applying that theorem again to this last relation and the third relation in (1) we obtain

(2) $(A \cup B) \cup C \subset A \cup (B \cup C)$.

It is clear that the names of the sets play no special role in (2), so that if we change the names we will obtain corresponding valid relations. In particular, if we interchange A and C in them, we obtain

(3) $$(C \cup B) \cup A \subset C \cup (B \cup A).$$

Applying Theorem 1.9 appropriately to (3) we see that

(4) $$A \cup (B \cup C) \subset (A \cup B) \cup C.$$

Putting (2) and (4) together yields (i).

(ii) From Theorems 1.10 and 1.9 we have

$$A \cap (B \cap C) \subset A, \quad B \cap C \subset B, \quad B \cap C \subset C,$$

and so applying the same theorems to these relations we obtain

(5) $\quad A \cap (B \cap C) \subset A, \quad A \cap (B \cap C) \subset B, \quad A \cap (B \cap C) \subset C.$

Using Theorem 1.12 on the relations (5) we conclude that

$$A \cap (B \cap C) \subset A \cap B;$$

applying the same theorem to this last relation and the third relation of (5) we derive

(6) $$A \cap (B \cap C) \subset (A \cap B) \subset C.$$

As in the proof of (i) we may clearly rename the sets in (6); we interchange A and C to obtain the corresponding relation

(7) $$C \cap (B \cap A) \subset (C \cap B) \cap A.$$

Now upon applying Theorem 1.9 appropriately to (7) we obtain

$$(A \cap B) \cap C \subset A \cap (B \cap C),$$

which together with (6) completes the proof of (ii).

The next theorem establishes a relationship between unions and intersections.

1.17 Theorem. If A, B, and C are sets, then

(i) $A \cap (B \cup C) = (A \cap B) \cup (A \cap C)$ and

(ii) $A \cup (B \cap C) = (A \cup B) \cap (A \cup C)$.

Proof. (i) We suppose, if possible, that $A \cap (B \cup C) \not\subset (A \cap B) \cup (A \cap C)$. Then there exists $x \in A \cap (B \cup C)$ such that $x \notin (A \cap B) \cup (A \cap C)$, that is, $x \in A$, and $x \in (B \cup C)$, but $x \notin A \cap B$ and $x \notin A \cap C$. These last two conditions, with $x \in A$, require that $x \notin B$ and $x \notin C$, that is, $x \notin B \cup C$. This contradiction proves that

(1) $$A \cap (B \cup C) \subset (A \cap B) \cup (A \cap C).$$

Since $B \subset B \cup C$ and $C \subset B \cup C$, then by Theorem 1.14 $A \cap B \subset A \cap (B \cup C)$ and $A \cap C \subset A \cap (B \cup C)$, whence by Theorem 1.12, $(A \cap B) \cup (A \cap C) \subset A \cap (B \cup C)$. Combining this with (1) completes the proof of (i).

(ii) This time we suppose, if possible, that

$$(A \cup B) \cap (A \cup C) \not\subset A \cup (B \cap C).$$

Then there exists $x \in (A \cup B) \cap (A \cup C)$ such that $x \notin A \cup (B \cap C)$, that is, $x \in A \cup B$ and $x \in A \cup C$ but $x \notin A$ and $x \notin B \cap C$. The first two conditions together with $x \notin A$ require that $x \in B$ and $x \in C$, that is, $x \in B \cap C$. This contradiction establishes

(2) $$(A \cup B) \cap (A \cup C) \subset A \cup (B \cap C).$$

Since $B \cap C \subset B$ and $B \cap C \subset C$, then by Theorem 1.14, $A \cup (B \cap C) \subset A \cup B$ and $A \cup (B \cap C) \subset A \cup C$, whence by Theorem 1.12, $A \cup (B \cap C) \subset (A \cup B) \cap (A \cup C)$. Combining this with (2) completes our proof.[10]

1.18 Definition. If A and B are sets, they are said to be *disjoint* if and only if $A \cap B = \varnothing$. If $\mathscr{F}$ is a family of sets, it is said to be *disjoint* if and only if $A \cap B = \varnothing$ whenever $A \in \mathscr{F}$, $B \in \mathscr{F}$, and $A \neq B$.

1.19 Definition. A set A is called a *singleton* if and only if there exists x such that $A = \{t \mid t = x\}$. We denote $\{t \mid t = x\}$ by $\langle x \rangle$.[11]

It is obvious that $x \in \langle x \rangle$ and if $z \in \langle x \rangle$ then $z = x$, that is, $z \in \langle x \rangle$ if and only if $z = x$. It is clear that we may repeat this operation defining a singleton to obtain $\langle (\langle x \rangle) \rangle = \{t \mid t = \langle x \rangle\}$. For convenience we shall omit the parentheses in such cases, and agree that $\langle (\langle x \rangle) \rangle = \langle \langle x \rangle \rangle$.

The reader is advised to keep in mind the distinction between x and $\langle x \rangle$; they are quite different things (see Exercise 55 at the end of this chapter).

We gave definitions above that allow us to consider the unions and intersections of two sets at a time, and it is possible to extend these concepts to permit the formation of unions and intersections of three or more sets by repeated application of the definitions already given. Furthermore, it is

[10] If we use $+$ and $\cdot$ to denote $\cup$ and $\cap$, respectively, this theorem says that $A \cdot (B + C) = A \cdot B + A \cdot C$ and $A + (B \cdot C) = (A + B) \cdot (A + C)$. The first of these is true of the corresponding algebraic operations of addition and multiplication, but the second certainly is not.

[11] In more formal treatments than this a somewhat different definition of singletons is given, which reduces to ours whenever x is a set.

also possible to show that in forming the union or intersection of several sets, the result does not depend on the order nor the grouping of the sets concerned. These things are consequences of Theorems 1.9 and 1.16.[12] However, we wish to introduce definitions of unions and intersections for arbitrary collections of sets that reduce to those already given if the collections consist of two sets and that make no reference to the magnitude of the collections. This is necessary, since we have need to form unions and intersections of collections containing infinitely many members.[13] We now state these definitions.

1.20 Definition. If $\mathscr{F}$ is a family of sets, then

$$\bigcup \mathscr{F} = \{x \mid \text{there exists } A \text{ such that } A \in \mathscr{F} \text{ and } x \in A\}; \text{ and}$$
$$\bigcap \mathscr{F} = \{x \mid \text{for all } A, \text{ if } A \in \mathscr{F} \text{ then } x \in A\}.$$

$\bigcup \mathscr{F}$ and $\bigcap \mathscr{F}$ are called, respectively, the *union* and *intersection* of $\mathscr{F}$.

It is intuitively clear that $\bigcup \mathscr{F}$ is the set obtained by combining all objects belonging to the members of $\mathscr{F}$ into a single set; also $\bigcap \mathscr{F}$ is the set of all objects that are common to all the members of $\mathscr{F}$ if $\mathscr{F} \neq \varnothing$.

1.21 Theorem. If $\mathscr{F}$ is a family of sets and $A \in \mathscr{F}$, then $\bigcap \mathscr{F} \subset A$ and $A \subset \bigcup \mathscr{F}$.

Proof. We take an arbitrary element $x \in \bigcap \mathscr{F}$. Since $A \in \mathscr{F}$, it follows from Definition 1.20 that $x \in A$. Hence $\bigcap \mathscr{F} \subset A$.

Next, we take an arbitrary element $x \in A$. Since $x \in A$ and $A \in \mathscr{F}$, it follows that $x \in \bigcup \mathscr{F}$; hence $A \subset \bigcup \mathscr{F}$.

1.22 Theorem. If A is a set and $\mathscr{F} = \langle A \rangle$, then $\bigcap \mathscr{F} = A = \bigcup \mathscr{F}$.

Proof. Since clearly $A \in \mathscr{F}$, then due to Theorem 1.21 we need prove only that $A \subset \bigcap \mathscr{F}$ and $\bigcup \mathscr{F} \subset A$. We consider first an arbitrary element $x \in A$. We also take an arbitrary element $z \in \mathscr{F}$. Then $z \in \langle A \rangle$ and so $z = A$; consequently $x \in z$. Thus we see that for all z, $z \in \mathscr{F}$ implies $x \in z$, whence by Definition 1.20, $x \in \bigcap \mathscr{F}$. Therefore we infer that if $x \in A$, then $x \in \bigcap \mathscr{F}$, and so $A \subset \bigcap \mathscr{F}$.

Next we take an arbitrary element $x \in \bigcup \mathscr{F}$. By Definition 1.20 there exists z such that $x \in z$ and $z \in \mathscr{F}$. But then $z \in \langle A \rangle$, so that $z = A$. Consequently $x \in A$. From this we conclude that $\bigcup \mathscr{F} \subset A$.

1.23 Theorem. If $\mathscr{F}$ is a non-empty family of sets and $x = y$ whenever $x \in \mathscr{F}$ and $y \in \mathscr{F}$, then $\mathscr{F} = \langle \bigcap \mathscr{F} \rangle$ and conversely.

Proof. If $\mathscr{F} = \langle \bigcap \mathscr{F} \rangle$, then clearly if $x \in \mathscr{F}$ and $y \in \mathscr{F}$, we have $x = \bigcap \mathscr{F} = y$, which completes the proof of the converse statement.

---

[12] For this reason parentheses may be omitted when writing unions or intersections of finite numbers of sets. We shall often write, for example, $A \cup B \cup C$ and $A \cap B \cap C$.
[13] We will give a formal definition of infinite sets in Chapter 3.

We now assume $\mathscr{F} \neq \varnothing$ and that $x = y$ whenever $x \in \mathscr{F}$ and $y \in \mathscr{F}$. There exists $A \in \mathscr{F}$. For all x, if $x \in \langle A \rangle$ then $x = A$ and so $x \in \mathscr{F}$; consequently, $\langle A \rangle \subset \mathscr{F}$. Similarly, for all x, if $x \in \mathscr{F}$, then $x = A$ by our hypotheses concerning $\mathscr{F}$, and so $x \in \langle A \rangle$. Thus $\mathscr{F} \subset \langle A \rangle$. We see now that $\mathscr{F} = \langle A \rangle$. Applying Theorem 1.22, we obtain $\mathscr{F} = \langle \bigcap \mathscr{F} \rangle$, as required.

1.24   Theorem. If $\mathscr{G}$ and $\mathscr{H}$ are families of sets, then $\bigcup (\mathscr{G} \cup \mathscr{H}) = (\bigcup \mathscr{G}) \cup (\bigcup \mathscr{H})$.
Proof. We take an arbitrary element $x \in \bigcup (\mathscr{G} \cup \mathscr{H})$. There exists A such that $A \in (\mathscr{G} \cup \mathscr{H})$ and $x \in A$. If $A \in \mathscr{G}$, then clearly $x \in \bigcup \mathscr{G}$; if $A \in \mathscr{H}$, then similarly $x \in \bigcup \mathscr{H}$. In either case, then, $x \in ((\bigcup \mathscr{G}) \cup (\bigcup \mathscr{H}))$. Consequently, $\bigcup (\mathscr{G} \cup \mathscr{H}) \subset ((\bigcup \mathscr{G}) \cup (\bigcup \mathscr{H}))$.

If $x \in ((\bigcup \mathscr{G}) \cup (\bigcup \mathscr{H}))$, then $x \in \bigcup \mathscr{G}$ or $x \in \bigcup \mathscr{H}$. If $x \in \bigcup \mathscr{G}$, there exists $A \in \mathscr{G}$ such that $x \in A$. Clearly, $A \in (\mathscr{G} \cup \mathscr{H})$, so that $x \in \bigcup (\mathscr{G} \cup \mathscr{H})$. If $x \in \bigcup \mathscr{H}$, then there exists $B \in \mathscr{H}$ such that $x \in B$. But then $B \in (\mathscr{G} \cup \mathscr{H})$, so that $x \in \bigcup (\mathscr{G} \cup \mathscr{H})$. Hence, in any case, if $x \in ((\bigcup \mathscr{G}) \cup (\bigcup \mathscr{H}))$, then $x \in \bigcup (\mathscr{G} \cup \mathscr{H})$; therefore $((\bigcup \mathscr{G}) \cup (\bigcup \mathscr{H})) \subset \bigcup (\mathscr{G} \cup \mathscr{H})$. This completes the proof.

We turn next to the important concept of *ordered pair*. Shortly, we shall give a formal definition of this concept in terms of the set-theoretic ideas already introduced, but first we shall discuss what we are about to define and explain why we do what we do.

Every student of analytic geometry is familiar with expressions of the form (x, y), where x and y are real numbers. Such expressions are called *ordered pairs*. In analytic geometry they have the property of representing points in the plane, once a set of coordinate axes has been fixed. Two ordered pairs are equal if and only if they represent the same point; and so $(x, y) = (x', y')$ if and only if $x = x'$, $y = y'$. The ordered pair (x, y) is thus sensitive to the order or position of x and y; if these are interchanged, a distinct ordered pair (y, x) is obtained, unless $x = y$. In the theory of sets, we would like to have a similar but more general concept, in which x and y may be sets or other mathematical objects, and subject to the same rule of equality, namely $(x, y) = (x', y')$ if and only if $x = x'$ and $y = y'$. Of course, such general ordered pairs cannot any longer be associated with points of a geometric plane.

We could introduce the concept of ordered pair as a new mathematical entity consisting of a pair of parentheses with a comma between, and a vacant space on each side of the comma that may be filled in with essentially any mathematical objects or sets and then simply impose the desired equality condition axiomatically. If the reader chooses to take this point of view, then he may skip the next few paragraphs and resume reading after Theorem 1.26.

However, we prefer not to introduce new concepts completely on their own and without reference to previously developed ideas. Thus we seek a way of defining ordered pairs, with the equality property we wish, in terms of elementary set-theoretic ideas. Clearly, we cannot define the ordered pair (x, y) simply as the set with x and y as its only elements because the concept of membership was not endowed with the ability to distinguish any kind of order with respect to elements. Something a little more sophisticated is required, and we shall see that the following definition is suitable.

1.25 Definition. If x and y are sets or mathematical objects, then

$$(x, y) = \langle\langle x\rangle\rangle \cup \langle\langle x\rangle \cup \langle y\rangle\rangle.$$

We agree to call (x, y) an *ordered pair*, x its *first term*, y its *second term*.

1.26 Theorem. If (x, y) and (x', y') are ordered pairs, then $(x, y) = (x', y')$ if and only if $x = x'$ and $y = y'$.
Proof. We let $A = \langle\langle x\rangle\rangle \cup \langle\langle x\rangle \cup \langle y\rangle\rangle$ and $A' = \langle\langle x'\rangle\rangle \cup \langle\langle x'\rangle \cup \langle y'\rangle\rangle$. If $x = x'$ and $y = y'$, it is clear that $A = A'$, whence $(x, y) = (x', y')$.

Next, we suppose that $A = A'$. Since $\langle x\rangle \in A$, then $\langle x\rangle \in A'$; hence either $\langle x\rangle \in \langle\langle x'\rangle\rangle$ or $\langle x\rangle \in \langle\langle x'\rangle \cup \langle y'\rangle\rangle$. If the former holds, then $\langle x\rangle = \langle x'\rangle$ and $x = x'$. If the latter is true, then $\langle x\rangle = \langle x'\rangle \cup \langle y'\rangle$. Since x' belongs to the right side of this last equation, it is also a member of the left side; thus $x' \in \langle x\rangle$ and $x' = x$. Hence, in any case, if $A = A'$, then $x = x'$. Therefore

$$\langle\langle x\rangle\rangle \cup \langle\langle x\rangle \cup \langle y\rangle\rangle = A = A' = \langle\langle x\rangle\rangle \cup \langle\langle x\rangle \cup \langle y'\rangle\rangle.$$

Since $\langle x\rangle \cup \langle y\rangle$ is an element of A, it also belongs to A', and consequently

(1) $\qquad \langle x\rangle \cup \langle y\rangle = \langle x\rangle \quad$ or $\quad \langle x\rangle \cup \langle y\rangle = \langle x\rangle \cup \langle y'\rangle.$

Similarly, $\langle x\rangle \cup \langle y'\rangle$ belongs to A', and therefore to A; hence

(2) $\qquad \langle x\rangle \cup \langle y'\rangle = \langle x\rangle \quad$ or $\quad \langle x\rangle \cup \langle y'\rangle = \langle x\rangle \cup \langle y\rangle.$

Whichever of the relations in (1) holds, we see that $(\langle x\rangle \cup \langle y\rangle) \subset (\langle x\rangle \cup \langle y'\rangle)$. Since y is a member of the left side of this inclusion, it is a member of the right side. Hence either $y \in \langle x\rangle$ or $y \in \langle y'\rangle$, that is, either $y = x$ or $y = y'$. By an exactly analogous argument using (2), we also infer that $y' = x$ or $y' = y$. Consequently, $y' = y$ and the theorem is proved.

Now that we have introduced the concept of ordered pairs, we can construct ordered pairs whose first or second terms are themselves ordered pairs, and so on, using the set-builder. Every set of the form

$$\{z \mid \text{there exist } x \text{ and } y \text{ such that } z = (x, y) \text{ and } \underline{\hspace{3cm}}\},$$

where the blank space is filled in with an admissible statement involving x and y, is a set of ordered pairs. For convenience and brevity, when we wish to define a set of ordered pairs we shall adopt the convention of writing merely

$$\{(x, y) \mid \underline{\hspace{3cm}}\},$$

where the blank space is filled in with the admissible statement mentioned above.

1.27 Definition. A set of ordered pairs is called a *relation*.

1.28 Definition. If Q is a relation, then

$$\{x \mid \text{there exists } y \text{ such that } (x, y) \in Q\}$$

is called the *domain* of Q, written $\mathscr{D}(Q)$.

1.29 Definition. If Q is a relation, then

$$\{y \mid \text{there exists } x \text{ such that } (x, y) \in Q\}$$

is called the *range* of Q, written $\mathscr{R}(Q)$.

It should be clear that both variables x and y appearing in the definitions of $\mathscr{D}(Q)$ and $\mathscr{R}(Q)$ are dummy variables, so there is no special significance in the use of x to denote a typical element of $\mathscr{D}(Q)$ or of y to denote a typical element of $\mathscr{R}(Q)$. The important thing is that $\mathscr{D}(Q)$ is the set of first terms of ordered pairs in Q, and $\mathscr{R}(Q)$ is the set of second terms of ordered pairs in Q.

We now give an example of a relation. Consider

$$Q = \{(x, y) \mid x \text{ is a real number and } 0 \leq y \leq x^2\}.$$

If we plot this relation on a cartesian graph, it is intuitively clear that it consists of all points in the plane between the X-axis and the parabola $y = x^2$, including the points on the X-axis and the parabola. Intuitively, we also see that $\mathscr{D}(Q)$ consists of all real numbers and $\mathscr{R}(Q)$ is the set of all non-negative numbers. We will not stop to prove these things, however, for we wish only to illustrate the ideas with this example.

As a second example, consider

$$Q = \{(x, y) \mid x \text{ and } y \text{ are real numbers and } 4x^2 + 9y^2 \leq 36\}.$$

If we plot this relation graphically, it is intuitively obvious that it consists of all points in the plane inside and on the ellipse $4x^2 + 9y^2 = 36$. If x is any real number such that $-3 \leq x \leq 3$, then clearly $x^2 \leq 9$, $4x^2 + 0 \leq 36$, and so $(x, 0) \in Q$. Hence for each such x, there exists y such that $(x, y) \in Q$, and $x \in \mathscr{D}(Q)$. On the other hand, if $x < -3$ or $3 < x$, then $x^2 > 9$, $4x^2 > 36$, and there exists no real number y such that $4x^2 + 9y^2 \leq 36$; thus there exists no y such that $(x, y) \in Q$; consequently $x \notin \mathscr{D}(Q)$. We conclude that $\mathscr{D}(Q) = \{x \mid -3 \leq x \leq 3\}$. In similar fashion, we see that $\mathscr{R}(Q) = \{y \mid -2 \leq y \leq 2\}$.

We conclude these examples by considering

$$Q = \{(x, y) \mid x \subset A \text{ and } y = \varnothing \text{ or } y = A\},$$

where here we are assuming that A is a fixed non-empty set. If $x \in \mathscr{D}(Q)$, there exists y such that $(x, y) \in Q$, whence $x \subset A$. Thus $\mathscr{D}(Q) \subset \{x \mid x \subset A\}$. Also if $x \subset A$, then $(x, \varnothing) \in Q$, so there exists y such that $(x, y) \in Q$, and so $x \in \mathscr{D}(Q)$. We infer that $\{x \mid x \subset A\} \subset \mathscr{D}(Q)$, and conclude that $\mathscr{D}(Q) = \{x \mid x \subset A\}$. If $y \in \mathscr{R}(Q)$, there exists x such that $(x, y) \in Q$, thus $y = \varnothing$ or $y = A$, and so $y \subset (\langle \varnothing \rangle \cup \langle A \rangle)$. Hence $\mathscr{R}(Q) \subset (\langle \varnothing \rangle \cup \langle A \rangle)$. Now if $y \in (\langle \varnothing \rangle \cup \langle A \rangle)$, then $y = \varnothing$ or $y = A$, and since $(A, \varnothing) \in Q$ and $(A, A) \in Q$, then in either case $(A, y) \in Q$. Thus there exists x such that $(x, y) \in Q$, and consequently $y \in \mathscr{R}(Q)$. Therefore $(\langle \varnothing \rangle \cup \langle A \rangle) \subset \mathscr{R}(Q)$. Thus $\mathscr{R}(Q) = \langle \varnothing \rangle \cup \langle A \rangle$.

1.30 **Definition.** If Q is a relation, then $\{(x, y) \mid (y, x) \in Q\}$ is evidently a relation; it is called the *inverse* of Q, written Q^{-1}.

1.31 **Theorem.** If Q is a relation, then $(Q^{-1})^{-1} = Q$.
Proof. We take an arbitrary element $z \in Q$; there exist x and y such that $z = (x, y)$. By Definition 1.30, $(y, x) \in Q^{-1}$. Applying this definition to Q^{-1}, it follows that $(x, y) \in (Q^{-1})^{-1}$, and therefore $z \in (Q^{-1})^{-1}$. Hence $Q \subset (Q^{-1})^{-1}$. Next, we take an arbitrary element $z \in (Q^{-1})^{-1}$. There exist x and y such that $z = (x, y)$. By Definition 1.30, $(y, x) \in Q^{-1}$, and by a second application of this definition, $(x, y) \in Q$. Thus $z \in Q$, and we conclude that $(Q^{-1})^{-1} \subset Q$. This completes the proof.

1.32 **Theorem.** If Q is a relation, $A = \mathscr{D}(Q)$, and $B = \mathscr{R}(Q)$, then $A = \mathscr{R}(Q^{-1})$, and $B = \mathscr{D}(Q^{-1})$.
Proof. If x is an arbitrary element of A, there exists y such that $(x, y) \in Q$, whence $(y, x) \in Q^{-1}$, and so $x \in \mathscr{R}(Q^{-1})$. Thus $A \subset \mathscr{R}(Q^{-1})$. If z is an arbitrary element of $\mathscr{R}(Q^{-1})$, there exists w such that $(w, z) \in Q^{-1}$; then $(z, w) \in Q$, and so $z \in \mathscr{D}(Q) = A$. Thus $\mathscr{R}(Q^{-1}) \subset A$. We now see that $A = \mathscr{R}(Q^{-1})$. Applying to the relation Q^{-1} what we have just learned, and invoking Theorem 1.31, we infer that $\mathscr{D}(Q^{-1}) = \mathscr{R}((Q^{-1})^{-1}) = \mathscr{R}(Q) = B$.

1.33 **Theorem.** If Q and S are relations, and $Q \subset S$, then $\mathscr{D}(Q) \subset \mathscr{D}(S)$ and $\mathscr{R}(Q) \subset \mathscr{R}(S)$.

Proof. If $x \in \mathscr{D}(Q)$, there exists y such that $(x, y) \in Q$; since $Q \subset S$, then $(x, y) \in S$. Thus $x \in \mathscr{D}(S)$. Hence $\mathscr{D}(Q) \subset \mathscr{D}(S)$.

If $y \in \mathscr{R}(Q)$, there exists x such that $(x, y) \in Q$, but then $(x, y) \in S$, and so $y \in \mathscr{R}(S)$. Therefore $\mathscr{R}(Q) \subset \mathscr{R}(S)$.

1.34 Definition. If A and B are sets, then $A \times B = \{(x, y) \mid x \in A$ and $y \in B\}$. We agree to call $A \times B$ the *cartesian product* of A and B.

It is easily checked that if Q is a relation, then $Q \subset \mathscr{D}(Q) \times \mathscr{R}(Q)$.

We are now ready to introduce the concept of *function*, which is perhaps the most important single idea in mathematics.

1.35 Definition. A relation Q is called a *function* if and only if for all x, u, and v, if $(x, u) \in Q$ and $(x, v) \in Q$, then $u = v$.

A word about the significance of this definition is in order. If Q is any relation and $x \in \mathscr{D}(Q)$, there exists y such that $(x, y) \in Q$, that is, $\{y \mid (x, y) \in Q\} \neq \varnothing$ if $x \in \mathscr{D}(Q)$. In case Q is a function, we see from above that if $x \in \mathscr{D}(Q)$ and if u and v belong to $\{y \mid (x, y) \in Q\}$, then $u = v$. Thus $\{y \mid (x, y) \in Q\}$ is a singleton for each $x \in \mathscr{D}(Q)$. If for some $x \in \mathscr{D}(Q)$, the corresponding set $\{y \mid (x, y) \in Q\}$ is not a singleton, then Q is not a function. Under the definition we have given, there is no such thing as a multiple-valued function. For example, consider

$$Q = \{(x, y) \mid x \text{ and } y \text{ are real numbers and } x^2 + y^2 = 1\}.$$

It is not difficult to see that $\mathscr{D}(Q) = \{x \mid -1 \leq x \leq 1\}$, and that for an arbitrary member x of $\mathscr{D}(Q)$, the ordered pairs $(x, \sqrt{1 - x^2})$ and $(x, -\sqrt{1 - x^2})$ belong to Q. Since the second terms of these ordered pairs are unequal if $x \neq 1$ and $x \neq -1$, then Q is not a function according to our definition.

We can, however, define two functions Q_1 and Q_2 from Q by setting

$$Q_1 = \{(x, y) \mid (x, y) \in Q \text{ and } y \geq 0\}$$
$$Q_2 = \{(x, y) \mid (x, y) \in Q \text{ and } y \leq 0\}.$$

It is readily checked that Q_1 and Q_2 are indeed functions containing, respectively, all ordered pairs of the form $(x, \sqrt{1 - x^2})$ and $(x, -\sqrt{1 - x^2})$, where $x \in \mathscr{D}(Q)$. In this way the equation $x^2 + y^2 = 1$ may be thought of as defining not a two-valued function, but two separate functions. Certain other equations may be treated similarly.

1.36 Definition. If Q is a function and $x \in \mathscr{D}(Q)$, we shall denote the only member of $\{y \mid (x, y) \in Q\}$ by $Q(x)$.

At this point we observe that if we had started all our mathematics from the standpoint of set theory, in which case every mathematical entity would be a set, then we could define $Q(x)$ explicitly as

$$\bigcap(\{y \mid (x, y) \in Q\}),$$

since, by virtue of Theorem 1.23, this last would be precisely the only member of $\{y \mid (x, y) \in Q\}$.

We also note that if Q is a function, then our definition implies that $y = Q(x)$ if and only if $(x, y) \in Q$. Expressed in different language, $Q(x)$ is the one and only object that can serve as the second term in an ordered pair belonging to Q of which x is the first term.

We now take up a few examples of relations and their inverses, which we investigate to see whether or not they are functions.

To start with, we take

$$Q = \{z \mid z = (0, 3), \text{ or } z = (2, 1), \text{ or } z = (2, 0)\}.$$

Since $(2, 1) \in Q$ and $(2, 0) \in Q$, we see that Q is not a function because we have found two ordered pairs in Q whose first terms are equal but whose second terms are not; $\{y \mid (2, y) \in Q\}$ is not a singleton. Now

$$Q^{-1} = \{z \mid z = (3, 0) \text{ or } z = (1, 2), \text{ or } z = (0, 2)\}.$$

Clearly Q^{-1} is a function, since there do not exist any ordered pairs in Q^{-1} with the same first term and different second terms. We see also that $Q^{-1}(3) = 0$, $Q^{-1}(1) = 2$, and $Q^{-1}(0) = 2$.

We consider next $Q = \{(x, y) \mid x \text{ is a real number and } y = x^2\}$. We assume that $(x, u) \in Q$ and $(x, v) \in Q$. Due to the definition of Q, $u = x^2$ and $v = x^2$, whence $u = v$, and so Q is a function. Clearly, if x is a real number, then $Q(x) = x^2$, since $(x, x^2) \in Q$. Now

$$Q^{-1} = \{(x, y) \mid y \text{ is a real number and } x = y^2\}.$$

We note that $(1, 1) \in Q^{-1}$ and $(1, -1) \in Q^{-1}$, whence Q^{-1} fails to qualify as a function. Of course, it happens that for every ordered pair $(x, y) \in Q^{-1}$, there is a second ordered pair $(x, -y) \in Q$, if $x \neq 0$. However, we need to display only one specific example of this phenomenon in order to show that Q^{-1} is not a function.

It is customary to denote functions by f, g, h, etc., and we will often follow this practice.

1.37 Theorem. If f is a function, then

$$f = \{z \mid \text{there exists } u \text{ such that } u \in \mathscr{D}(f) \text{ and } z = (u, f(u))\}$$
$$\mathscr{R}(f) = \{v \mid \text{there exists } u \text{ such that } u \in \mathscr{D}(f) \text{ and } v = f(u)\}.$$

Proof. We let $A = \{z \mid \text{there exists } u \text{ such that } u \in \mathscr{D}(f) \text{ and } z = (u, f(u))\}$. $B = \{v \mid \text{there exists } u \text{ such that } u \in \mathscr{D}(f) \text{ and } v = f(u)\}$.

If z is an arbitrary member of f, then there exists an ordered pair $(x, y) = z \in f$, whence $y = f(x)$ by Definition 1.36. Thus $z = (x, f(x))$. Also $x \in \mathscr{D}(f)$, and so $z \in A$. Hence $f \subset A$.

If z is an arbitrary member of A, then there exists $x \in \mathscr{D}(f)$ such that $z = (x, f(x))$. Since $x \in \mathscr{D}(f)$, there exists y such that $(x, y) \in f$; and by Definition 1.36, $y = f(x)$, whence $z = (x, y) \in f$. Thus $A \subset f$. This completes the proof of the first part of the theorem.

We take an arbitrary element $y \in \mathscr{R}(f)$. There exists x such that $(x, y) \in f$; and $x \in \mathscr{D}(f)$. Thus by Definition 1.36, $y = f(x)$, so that $y \in B$. Hence $\mathscr{R}(f) \subset B$. Now if y is an arbitrary element of B, then there exists $x \in \mathscr{D}(f)$ such that $y = f(x)$. Since $x \in \mathscr{D}(f)$ there exists w such that $(x, w) \in f$; and therefore $w = f(x)$. But then $y = w$, so that $(x, y) \in f$ and $y \in \mathscr{R}(f)$. Hence $B \subset \mathscr{R}(f)$. This completes the proof of the theorem.

1.38 Definition. If f and g are functions, we say that g is an *extension* or *continuation* of f if and only if $f \subset g$.

1.39 Theorem. A necessary and sufficient condition that g be an extension of f is that $f(x) = g(x)$ for all $x \in \mathscr{D}(f)$.
Proof. If x is an arbitrary member of $\mathscr{D}(f)$, there exists y such that $(x, y) \in f$ and $y = f(x)$. If $f \subset g$, then $(x, y) \in g$ and $y = g(x)$. Thus $f(x) = g(x)$, as required.

Suppose next that for all $x \in \mathscr{D}(f)$, $f(x) = g(x)$. We take an arbitrary element $z \in f$. There exists x and y such that $z = (x, y)$. Now $x \in \mathscr{D}(f)$, so that $y = f(x) = g(x)$. Thus $(x, y) = (x, g(x)) \in g$ by Theorem 1.37, so that $z \in g$. Hence $f \subset g$ as required.

From the preceding theorem we obtain the following obvious result.

1.40 Corollary. If f and g are functions, then $f = g$ if and only if they have a common domain and $f(x) = g(x)$ whenever x is in that domain.

There are times when a function is defined on a set A, but one wishes only to consider its behavior on the common part of a certain set B with A. In these situations the *restriction* concept is useful.

1.41 Definition. If f is a function, $\mathscr{D}(f) = A$ and B is a set, then we define $(f \mid B)$, the *restriction of f to B*, as follows:

$$(f \mid B) = \{(x, y) \mid (x, y) \in f \text{ and } x \in B\}.$$

It is easily checked that $(f \mid B) \subset f$ and that $\mathscr{D}(f \mid B) = B \cap A$. From Theorem 1.39 it then follows that $(f \mid B)(x) = f(x)$ whenever $x \in B \cap A$.

If f is a function, then each $x \in \mathscr{D}(f)$ may be thought of as being operated upon by f to obtain a new entity $f(x)$. From this point of view x is transformed by f into $f(x)$. This may serve to motivate the following definition.

1.42 **Definition.** If f is a function, $A = \mathscr{D}(f)$, $B = \mathscr{R}(f) \subset C$, then we say that A is *transformed into*, or *mapped into*, or *carried into* C *under f*. If $C = B$, we replace the word "into" in each of its occurrences by "onto." If $x \in A$, then $f(x)$ is called the *image of x under f*. If $y \in B$, and x is any element of A such that $y = f(x)$, then x is called an *inverse image of y under f*. If $A' \subset A$ and $B' \subset B$, then we define

$$f(A') = \{y \mid y = f(x) \text{ for some } x \in A'\}$$

$$f^{-1}(B') = \{x \mid f(x) = y \text{ for some } y \in B'\}.$$

In the notation of the preceding definition, we see that if $x \in A$, then there is just one image of x in B under f, namely $f(x)$. Also if $y \in B$, there exists x in A such that $(x, y) \in f$, whence $x \in A$ and $y = f(x)$. Thus each element of B has some inverse image in A under f. However, there may be many such inverse images. A particularly important situation arises when there exists exactly one inverse image for each element of B, that is, the set of inverse images of y is a singleton for each $y \in B$. We use a special terminology to describe this situation.

1.43 **Definition.** If f is a function, it is said to be a *one-to-one* or a *univalent* mapping of $\mathscr{D}(f)$ onto $\mathscr{R}(f)$ if and only if $f(u) \neq f(v)$ whenever $u \in \mathscr{D}(f)$, $v \in \mathscr{D}(f)$ and $u \neq v$.

If f is a function, then f^{-1} is surely a relation, but it may not be a function. The next theorem gives a criterion ensuring that f^{-1} is a function.

1.44 **Theorem.** If f is a function, then f^{-1} is a function if and only if f is a one-to-one mapping of $\mathscr{D}(f)$ onto $\mathscr{R}(f)$.
Proof. We let $\mathscr{D}(f) = A$, $\mathscr{R}(f) = B$. By Theorem 1.32, $A = \mathscr{R}(f^{-1})$ and $B = \mathscr{D}(f^{-1})$.

We assume first that f is a one-to-one mapping of A onto B. We consider arbitrary y, u, and v such that $(y, u) \in f^{-1}$ and $(y, v) \in f^{-1}$. Then $(u, y) \in f$ and $(v, y) \in f$, whence $f(u) = y = f(v)$. By Definition 1.43 it follows that $u = v$, and so by Definition 1.35, f^{-1} is a function.

Next, we assume that f^{-1} is a function. We take arbitrary members u and v of the set A, such that $u \neq v$, and we let $w = f(u)$, $z = f(v)$. Now $(u, w) \in f$ and $(v, z) \in f$, so that $(w, u) \in f^{-1}$ and $(z, v) \in f^{-1}$. Thus $u = f^{-1}(w)$, and $v = f^{-1}(z)$. Since $u \neq v$, these last equations force us to conclude that $w \neq z$. Thus $f(u) \neq f(v)$, and consequently f is a one-to-one mapping of A onto B.

1.45 **Corollary.** If f is a one-to-one mapping, then f^{-1} is a one-to-one mapping of $B = \mathscr{R}(f)$ onto $A = \mathscr{D}(f)$; furthermore

$$f^{-1}(f(x)) = x \text{ for each } x \in A, \text{ and } f(f^{-1}(y)) = y \text{ for each } y \in B.$$

Proof. Theorem 1.44 guarantees that f^{-1} is a function; the same theorem says that $(f^{-1})^{-1} = f$ can be a function only if f^{-1} is a one-to-one mapping.

If $x \in A$, then $(x, f(x)) \in f$ and so $(f(x), x) \in f^{-1}$; hence $f^{-1}(f(x)) = x$; if $y \in B$, then $(y, f^{-1}(y)) \in f^{-1}$ and so $(f^{-1}(y), y) \in f$; thus $f(f^{-1}(y)) = y$.

1.46 **Theorem.** If f and g are such functions that $\mathscr{R}(f) \subset \mathscr{D}(g)$ and

$$Q = \{(x, y) \mid x \in \mathscr{D}(f) \quad \text{and} \quad y = g(f(x))\},$$

then Q is a function, $\mathscr{D}(Q) = \mathscr{D}(f)$, and $\mathscr{R}(Q) \subset \mathscr{R}(g)$.

Proof. If (x, u) and (x, v) are arbitrary ordered pairs belonging to Q, then $x \in \mathscr{D}(f)$ and so $f(x) \in \mathscr{R}(f)$ by Theorem 1.37. Since $\mathscr{R}(f) \subset \mathscr{D}(g)$, then $f(x) \in \mathscr{D}(g)$, so that $g(f(x))$ has a definite meaning in accordance with Definition 1.36. Now $u = g(f(x)) = v$, and therefore Q is a function.

Clearly, if $x \in \mathscr{D}(f)$, then $(x, g(f(x)) \in Q$, so that $x \in \mathscr{D}(Q)$, and consequently $\mathscr{D}(f) \subset \mathscr{D}(Q)$. If $x \in \mathscr{D}(Q)$, then there exists y such that $(x, y) \in Q$, consequently $x \in \mathscr{D}(f)$, and so $\mathscr{D}(Q) \subset \mathscr{D}(f)$. Thus $\mathscr{D}(Q) = \mathscr{D}(f)$. If $y \in \mathscr{R}(Q)$, there exists x such that $x \in \mathscr{D}(f), y = g(f(x))$, and so $y \in \mathscr{R}(g)$ due to Theorem 1.37. Hence $\mathscr{R}(Q) \subset \mathscr{R}(g)$.

The preceding theorem gives point to the following definition.

1.47 **Definition.** If f and g are such functions that $\mathscr{R}(f) \subset \mathscr{D}(g)$ then we agree that the function

$$\{(x, y) \mid x \in \mathscr{D}(f) \quad \text{and} \quad y = g(f(x))\}$$

shall be called the *composition of g with f*, and we shall denote it by $g \circ f$.

1.48 **Theorem.** If f and g are functions, $A = \mathscr{D}(f)$, $B = \mathscr{R}(f) = \mathscr{D}(g)$, and $\mathscr{R}(g) = C$, then $g \circ f$ maps A onto C. Furthermore, if f and g are one-to-one mappings of their domains onto their respective ranges, then so is $g \circ f$.

Proof. We let $h = g \circ f$ for convenience. By Theorem 1.46, $\mathscr{D}(h) = A$ and $\mathscr{R}(h) \subset C$. Now we take an arbitrary element $z \in C$. There exists y such that $(y, z) \in g$, so that $y \in \mathscr{D}(g) = \mathscr{R}(f)$ and $z = g(y)$. Then there exists x such that $(x, y) \in f$, and so $x \in \mathscr{D}(f) = A$ and $y = f(x)$. Thus $z = g(y) = g(f(x)) = h(x)$, whence $z \in \mathscr{R}(h)$ by Theorem 1.37. We conclude that $C \subset \mathscr{R}(h)$, and finally $\mathscr{R}(h) = C$. This shows that $h = g \circ f$ maps A onto C.

We now assume that f and g are one-to-one mappings. We take arbitrary members u and v of $\mathscr{D}(h) = A$ such that $u \neq v$. Then $f(u) \neq f(v)$ and $h(u) = g(f(u)) \neq g(f(v)) = h(v)$ since g is a one-to-one mapping. This shows that $h = g \circ f$ is a one-to-one mapping.

1.49 **Definition.** If $\mathscr{F}$ is a family of sets, then $\mathscr{F}$ is said to be a *nest* if and only if whenever $A \in \mathscr{F}$ and $B \in \mathscr{F}$, then either $A \subset B$ or $B \subset A$.

1.50 Theorem. If $\mathcal{F}$ is a family of functions (that is, the members of $\mathcal{F}$ are functions) and $\mathcal{F}$ is a nest, then $\bigcup \mathcal{F}$ is a function.

Proof. We let $F = \bigcup \mathcal{F}$. If $z \in F$, then there exists $f \in \mathcal{F}$ such that $z \in f$. Since f is a function, then z is an ordered pair. Thus F is a relation.

We now consider arbitrary ordered pairs (x, u) and (x, v) belonging to F. There exist functions f and g belonging to $\mathcal{F}$ such that $(x, u) \in f$ and $(x, v) \in g$. Also, either $f \subset g$ or $g \subset f$ since $\mathcal{F}$ is a nest. If $f \subset g$, then $(x, u) \in g$. But also $(x, v) \in g$, and since g is a function, we must have $u = v$. If $g \subset f$, then $(x, v) \in f$. But $(x, u) \in f$, and since f is a function, then $u = v$. Hence in either case $u = v$, which means F is a function.

This completes our study of functions in general. Before dropping the subject, we note that Corollary 1.40 makes possible a kind of abbreviation that is frequently used by mathematicians, and it is one we will frequently use in the future. To explain what we have in mind, suppose we wish to define a function. We may do this by specifying the ordered pairs that belong to it; but according to Corollary 1.40, we achieve the same result by specifying the domain and giving the value of $f(x)$ for an arbitrary member x of the domain, since functions having exactly the same values are necessarily the same. This technique is used in the following exercises, and often afterward. However, it is essential to tell just what the domain is to be in such cases; otherwise, in effect, the function is not properly defined.

1.51 The Axiom of Choice. The foregoing material forms the basis for those portions of set theory that we shall require—with one noteworthy exception that we will introduce shortly. To prepare for this, we need first to discuss certain ideas related to axiomatic systems in general, and those in particular which are devised to represent formal models of intuitive concepts. In the latter cases, the usual procedure consists of writing in formal symbolic language exactly those fundamental properties that seem to characterize the intuitive ideas in question. It must be stressed that these fundamental properties are not forced on us by strictly logical considerations; they are somewhat arbitrary and represent simply a choice that we deem appropriate. For this reason, the formal expressions of those properties that we ultimately adopt have to be regarded as axioms, that is, conditions imposed and not subject to proof. Additional properties may be discovered from the axioms using only logical inference. It is also permissible to define other concepts in terms of the basic ones and to prove properties of such defined concepts with the help of the axioms.

In such a logical development certain theorems will emerge that are in conformity with the intuitive understanding we have of the subject lying behind the formal system. The more such theorems we prove, the greater becomes our confidence in the validity of the formal theory as a model of the intuitive situation. If we prove a theorem that violates our intuition, then we re-examine the formal structure carefully to see whether it really incorporates the essence of the intuitive concepts, and we may decide to revise the formal system accordingly. Sometimes, however, it turns out that there is really nothing wrong; it may just happen that the intuition itself is too vague in some respects, and in these cases we accept the consequences of our reasoning, surprising though they may be.

However, we always discard a formal structure if we find logical inconsistencies in it, that is, if we are able to prove two logically contradictory statements from it. In some mathematical theories, including in particular the theory of sets, no inconsistencies have yet been found, but it is still unknown whether some lie hidden. Despite this possibility, we have considerable faith in the theory of sets, whose logical consequences include virtually all of mathematics, since it seems reasonable that in all the work that has been done over the years, contradictions would have been found if they were inherent in the system. Nevertheless the possibility of inconsistencies is still there. It is a difficult matter to show the impossibility for inconsistencies to exist in a system of axioms.

We come now to a specific connection between the foregoing discussion and our development of set theory. If A is a set, then the statement "$A \neq \varnothing$" is equivalent to "there exists x such that $x \in A$." This last is purely a formal expression, but it gives one the feeling that it should be possible to reach into A and actually pick out a member. However, the formal existence of an element in a set and the act of selection are different matters entirely. The selection of a member from some non-empty sets presents no problem, for in some cases one can construct a member, that is, write an expression for an explicit member. For example, consider two real numbers y and z with $y < z$, and consider $\{x \mid y < x < z\}$. It seems clear that there exist real numbers between y and z, but without being given any more information than we already have, is it possible to construct a number in this set? In this case, the answer is in the affirmative, since the arithmetic mean of y and z, namely $(y + z)/2$, is an expression for a real number that can easily be shown to lie strictly between y and z and so belongs to the set under consideration. Other sets may be defined in such a complex manner that although it may be possible to show indirectly that they are not empty, it may still not be known how to construct a member of them, or indeed whether such a construction exists.

This problem becomes critical in the following situation. Consider a non-empty family $\mathscr{F}$ of non-empty sets. It puts no strain on the intuition to imagine a set E made up by selecting exactly one element from each of the sets belonging to $\mathscr{F}$. As it happens, the proofs of certain important theorems in mathematics depend on assuming the existence of such a set E; at least, no other way of proving them is known. At the same time, no one succeeded in proving the existence of E, starting from a perfectly general family $\mathscr{F}$ and using only logical inference on the axioms of set theory (which, we remind the reader, were not given fully herein for reasons mentioned at the outset). This state of affairs led some mathematicians to assert the existence of E as still another axiom of set theory, called the *axiom of choice*.

One might well ask what justification and what hazards are involved in proclaiming the above an axiom. It could conceivably happen that the statement of the axiom of choice would turn out to be a theorem of the other axioms of set theory, in which case no real harm would result; mathematicians would have been guilty only of using a redundant axiom. The only real danger would lie in the possibility that, in adding the axiom of choice to the other axioms, inconsistencies might be introduced into the system that were not there previously. However, this possibility was ruled out by the work of K. Gödel; although we do not know if the other axioms of set theory form a consistent system, he proved that adding the axiom of choice to them does not change whatever consistency properties they had beforehand.

The whole question was settled shortly before this book went to press. P. J. Cohen of Stanford University established that it is utterly impossible to prove the axiom of choice from the other axioms of set theory, and so it is proper to call it an axiom. However, there are still some mathematicians who refuse to use the axiom of choice in their work. They will not acknowledge the legitimacy of proofs that depend on non-constructive methods, and since the axiom of choice is non-constructive in nature, they eschew it and devote themselves to seeing how far they can extend mathematics by constructive methods.

There are several forms of the axiom in common use. Without going into detail, it suffices to say that they all assume or have as a consequence the existence of a formal mechanism that picks out an element from any given non-empty set. We shall assume the following form of it: if $\mathscr{F}$ is any given non-empty family of non-empty sets, then there exists a function f with $\mathscr{D}(f) = \mathscr{F}$, such that $f(A) \in A$ whenever $A \in \mathscr{F}$. It is clear that f picks out an element from each member of $\mathscr{F}$. A special case arises if we have a function F whose domain is a non-empty set S, such that $F(x)$ is a non-empty set for each $x \in S$. We let $\mathscr{F} = \mathscr{R}(F)$. Clearly $\mathscr{F}$ is a non-empty family of non-empty sets; hence by our assumption there exists a function f with $\mathscr{D}(f) = \mathscr{R}(F)$ such that $f(A) \in A$ whenever $A \in \mathscr{R}(F)$. Now consider the composite function $g = f \circ F$, whose domain and range are S and $\mathscr{R}(f)$ respectively, by virtue of Theorem 1.48.

For any $x \in S$, we have $F(x) \in \mathcal{R}(F) = \mathcal{D}(f)$, and hence $g(x) = f(F(x)) \in F(x)$. In brief, we can make the following statement: there exists a function g with $\mathcal{D}(g) = \mathcal{D}(F)$ such that $g(x) \in F(x)$ whenever $x \in \mathcal{D}(F)$. We shall have occasion to use this form of the axiom of choice.

This completes our discussion of general set theory. From time to time we will apply and extend the foregoing general theory to certain specific situations.

Exercises

In the following problems the student is expected to know and make use of elementary properties of inequalities and integers.
1. Given $\{x \mid x = 1 \text{ or } x = 2 \text{ or } x = 3\}$, show that 1, 2, and 3 are members of this set.[12] Also show that if $z \neq 1$, $z \neq 2$, and $z \neq 3$, then z is not a member of the given set. Can you now list all the members of this set?
2. Given $\{x \mid x \text{ is a positive integer and } x \leq 11\}$, determine exactly what are the members of this set. List them.
3. Given $\{x \mid \text{there exists a positive integer } n \text{ such that } x = 2n\}$, prove that 2, 4, and 6 are members but that 7 is not. Can you describe in a simple sentence what the members of this set are?
4. Given $\{x \mid x \text{ is a positive integer and } x^2 \leq 17\}$, determine explicitly the members of this set.
5. Given $\{x \mid x \text{ is a real number and } 7x^3 - 3x - 1 < 0\}$, determine whether or not -1, 1, $1\frac{1}{2}$ are members.
In each of the following, write in brace expression form the following sets:
6. The set with 0 as its only member.
7. The set with -1 and $+1$ as its only members.
8. The set with x, y, and z as its only members.
9. The set with the first fifty positive integers as members.
10. The set whose members consist of all squares of positive integers.
11. The set of solutions of the equation $5x^6 - 2x^4 - x + 10 = 0$.
12. The set of all reciprocals of negative integers.
13. The set of all positive integral powers of 2.
14. The set of all positive integral powers of all positive integers.
15. Consider $\{x \mid x = 0 \text{ and } x = 1\}$. Show that if x is any member of this set, then $0 = 1$. What does this suggest to you concerning the membership of this set? (A discussion of this situation may be found following Definition 1.4.)
16. Given $\{x \mid x \text{ is a real number and } x^2 \leq 50\}$, show that 7, -3.2, -7, and 3.2 are members of this set. Show in general that if z is a member, then so is $-z$; and if z is not a member, then $-z$ is not a member.
17. Given $\{x \mid x \text{ is an even positive integer}\}$, show that if z is a member of this set, then so is kz, where k is any given positive integer. (*Note:* An even positive integer is one that is divisible by 2).
18. Given $\{x \mid x \text{ is a positive integer, } x \text{ is divisible by 2, and } x \text{ is divisible by 3}\}$, show that every positive integer divisible by 6 is a member of this set.
19. Given $\{x \mid x > t\}$, prove that if z is a member of this set and $s > z$ then s is a member. Is t a member? If $w < t$, is w a member?

---

[12] Follow the procedure explained on page 4.

20. Given the set $\{x \mid x = s \text{ or } x = t\}$ and assuming $s \neq t$, state whether or not s and t are members of this set and prove your statement.

21. Given the set $\{x \mid \text{for all } z, \text{ if } z > 2, \text{ then } z > x\}$,
    (a) show that 2 is a member of this set.
    (b) show that if t is any number less than 2, then t is a member of this set.
    (c) show that if t is any number greater than 2, then t is not a member of this set.

22. Show that the real number $2 - \sqrt{3}$ is not transcendental by actually finding an algebraic equation of which it is a root.

§§ 1.1 to 1.5

In the following problems you may assume and use elementary facts of algebra.

23. Prove that $\varnothing \subset A$, where A is an arbitrary set, by assuming the contrary and showing that a contradiction arises.

24. If $A = \{x \mid x \text{ is the square of some positive integer}\}$ and $B = \{x \mid x \text{ is the cube of some positive integer}\}$, prove that $A \not\subset B$ and $B \not\subset A$. Show that $64 \in A$ and $64 \in B$. Find a few other numbers belonging to both sets.

25. Given $A = \{x \mid x \text{ is the square of some positive integer}\}$, $B = \{x \mid x \text{ is the fourth power of some positive integer}\}$, prove that $A \supset B$ and $B \neq A$ (that is, B is a proper subset of A).

26. Let $A = \{x \mid x^2(x - 1)(x - 2)(x - 3) = 0\}$, $B = \{x \mid x = 0 \text{ or } x = 1 \text{ or } x = 2 \text{ or } x = 3\}$. Prove that $A = B$.

27. Given the real numbers s and t, with $s < t$, and the sets $A = \{x \mid x < s\}$, $B = \{x \mid x < t\}$, prove that $A \subset B$. Determine whether the real number $(s + t)/2$ belongs to A or B. On the basis of this decide whether $B \subset A$ or not.

28. Given the sets A, B, C satisfying the relations $A \subset C$, $B \subset C$, either prove that these relations imply that $A \subset B$ or else given an example of three such sets for which the statement is false.

29. Given sets A, B, C such that $A \subset B$ and $B \not\subset C$, either prove that necessarily $A \not\subset C$ or else give specific examples of sets A, B, C satisfying the first two but not the third condition.

30. Given the sets A, B, and C such that $A \subset B$ and $A \subset C$, either prove that these relations imply $B \not\subset C$ or else give an example of three such sets for which $B \subset C$.

31. Suppose $A = \{x \mid x \text{ is an even positive integer}\}$, $B = \{x \mid x \text{ is a positive integer divisible by 6}\}$. Prove that $B \subset A$ and $B \neq A$.

32. Given $A = \{x \mid x^7 - 2x^3 + 3x + 1 = 0\}$, $B = \{x \mid x^8 - 2x^4 + 3x^2 + x = 0\}$, show that $A \subset B$ and $A \neq B$.

33. Given $A = \{x \mid x = s \text{ or } x = t\}$, $B = \{x \mid x = s\}$, and $s \neq t$, determine what, if any, relationship exists between A and B. Actually determine all members of the two sets.

34. Let $A = \{x \mid x \text{ is an odd positive integer}\}$

    $B = \{x \mid x \text{ is an even positive integer}\}$

    $C = \{x \mid x = A \text{ or } x = B\}$.

    (a) What are the members of C?
    (b) Is it true or not that $A \subset C$? $B \subset C$? Give reasons for your conclusion.
    (c) Is it true that $1 \in C$? $2 \in C$? Give your reasons.

35. With reference to Problem 34 prove that

$$\{x \mid x \text{ is a positive integer}\} = \{x \mid x \text{ belongs to a member of } C\}.$$

36. If $A = \{x \mid x \neq s \text{ or } x \neq t\}$ and $s \neq t$, determine whether or not s and t are members of A. Write an expression for the set of all objects that do not belong to A. Justify your expression.

37. Let $A = \{t \mid t \in x \text{ or } t = x\}$. Prove that (a) $x \in A$ and (b) $x \subset A$.

38. Give a formal proof that if $\varnothing'$ is any set such that the statement "for all x, $x \notin \varnothing'$" is true, then $\varnothing' = \varnothing$. Do it first directly and then by the indirect method, that is, by assuming $\varnothing' \neq \varnothing$ and showing that a contradiction results.

§§ 1.6 to 1.19

In the Exercises 39 to 58, either prove that the stated relation is true for all sets A, B, C, etc. or else show by exhibiting suitable specific examples of these sets that the statement is not always true.

39. $(A - B) \subset A$.
40. If $A \cup C = B \cup C$, then $A = B$.
41. If $A \cap C = B \cap C$, then $A = B$.
42. $A \cap B \subset A \cup B$.
43. $A \subset B$ if and only if $A \cap B = A$.
44. $A \subset B$ if and only if $A \cup B = B$.
45. If $A \cap B \subset C$ then $A \subset C$ or $B \subset C$.
46. $(A - C) \cup (B - C) = (A \cup B) - C$.
47. $A - B = A$ if and only if A and B are disjoint.
48. $(A - B) \cup (B - A) = A \cup B$.
49. $A \cap B \cap C = (A - (A - B)) - C$.
50. $(A - B) \cup (B - C) = A - C$.
51. $(A - B) \cup (A \cap B) \cup (B - A) = A \cup B$.
52. $(A \cup B) \cup (C \cup D) = (D \cup A) \cup (C \cup B)$.
53. If $A - B = C$, then $A = B \cup C$.
54. If $A \subset B \cup C$ then $A \subset B$ or $A \subset C$.
55. Show that $x \neq \langle x \rangle$ for all x. *Hint:* If $x = \langle x \rangle$, then show that x must be a member of x, which is impossible.
56. If $x \in A$, then $\langle x \rangle \subset A$.
57. If $A \subset C$, $B \subset C$, then $C - (A \cup B) = (C - A) \cap (C - B)$ and $C - (A \cap B) = (C - A) \cup (C - B)$.
58. If $A \subset C$ and $B \subset C$ then $A = \varnothing$ if and only if $B = (A \cap (C - B)) \cup ((C - A) \cap B)$.
59. Given $A = \{x \mid -1 < x < 3\}$,
    $$B = \{x \mid 2 < x < 5\}.$$
    prove whether or not the following statements are true:
    (a) $A - B = \{x \mid -1 < x \leq 2\}$.
    (b) $A \cup B = \{x \mid -1 \leq x < 5\}$.
60. The *symmetric difference* Δ is defined with respect to pairs of sets A and B as follows: $A \Delta B = (A - B) \cup (B - A)$. Prove that (a) $A \Delta A = \varnothing$, (b) $A \Delta B = B \Delta A$, (c) $A \Delta (B \Delta C) = (A \Delta B) \Delta C$, (d) for any sets A and B there exists a unique set X such that $A \Delta X = B$.

§§ 1.20 to 1.24

61. If A and B are sets and $\mathscr{F} = \{x \mid x = A \text{ or } x = B\}$, then $\bigcup \mathscr{F} = A \cup B$ and $\bigcap \mathscr{F} = A \cap B$.
62. If $\mathscr{F} = \varnothing$ then $\bigcup \mathscr{F} = \varnothing$.
63. If $\mathscr{G}$ and $\mathscr{H}$ are families of sets such that $\mathscr{G} \subset \mathscr{H}$ then $\bigcup \mathscr{G} \subset \bigcup \mathscr{H}$ and $\bigcap \mathscr{H} \subset \bigcap \mathscr{G}$.
64. If $\mathscr{G}$ and $\mathscr{H}$ are families of sets, then $\bigcap (\mathscr{G} \cup \mathscr{H}) = (\bigcap \mathscr{G}) \cap (\bigcap \mathscr{H})$.
65. If A is a set, $\mathscr{G}$ is a family of sets, and $\mathscr{H} = \{x \mid \text{there exists } B \text{ such that } B \in \mathscr{G} \text{ and } x = A \cup B\}$, then $\bigcup \mathscr{H} = A \cup (\bigcup \mathscr{G})$ and $\bigcap \mathscr{H} = A \cup (\bigcap \mathscr{G})$.
66. If $\mathscr{F}$ is a family of sets and there exist $A \in \mathscr{F}$ and $B \in \mathscr{F}$ such that $A \cap B = \varnothing$, then $\bigcap \mathscr{F} = \varnothing$.
67. For any positive real number x, let us define $I_x = \{t \mid x < t < 2x\}$, and let $\mathscr{F} = \{A \mid \text{there exists } x > 0 \text{ such that } A = I_x\}$. Show that $\bigcup \mathscr{F}$ is the set of all positive real numbers and $\bigcap \mathscr{F} = \varnothing$.
68. For any positive real number x, let $J_x = \{t \mid -1/x < t < x\}$, and $\mathscr{G} = \{A \mid \text{there exists } x > 0 \text{ such that } A = J_x\}$. Determine logically what the sets $\bigcup \mathscr{G}$ and $\bigcap \mathscr{G}$ consist of.
69. Determine logically all possible solutions $\mathscr{F}$ of the equation $\bigcup \mathscr{F} = \varnothing$.
70. Given a set A, let $\mathscr{F} = \{B \mid B \subset A\}$. Show that if $\mathscr{G} \subset \mathscr{F}$, then $\bigcup \mathscr{G} \in \mathscr{F}$. Determine what the sets $\bigcup \mathscr{F}$ and $\bigcap \mathscr{F}$ consist of.
71. If $\mathscr{F}$ is a family of sets, and B is such a set that $A \subset B$ whenever $A \in \mathscr{F}$, then $\bigcup \mathscr{F} \subset B$; if C is such a set that $C \subset A$ whenever $A \in \mathscr{F}$, then $C \subset \bigcap \mathscr{F}$.

§§ 1.25 to 1.36

In Exercises 72 to 80 determine whether or not Q and Q^{-1} are functions. To show the contrary, actually exhibit ordered pairs with the same first term and different second terms. Find $\mathscr{D}(Q)$ and $\mathscr{R}(Q)$ in each case. Also write expressions for $Q(x)$ and $Q^{-1}(x)$ where these are defined. You may use any methods of elementary algebra known to you in connection with these problems.

72. $Q = \{(x, y) \mid x \text{ and } y \text{ are real numbers and } 4x^2 + y^2 = 4\}$.
73. $Q = \{(x, y) \mid x \text{ and } y \text{ are real numbers and } x^4 = y^2\}$.
74. $Q = \{(x, y) \mid -1 \leq x \leq 2 \text{ and } y = x^2 - 4x\}$.
75. $Q = \{(x, y) \mid 0 \leq x \leq 2 \text{ and } y = 3 |x|\}$.
76. $Q = \{(x, y) \mid x \text{ and } y \text{ are real numbers and } x = 1 \text{ or } y = 1\}$.
77. $Q = \{(x, y) \mid x \text{ is a positive integer and } y = 2^x\}$.
78. $Q = \{(x, y) \mid -1 \leq x \leq 3; \text{ if } -1 \leq x \leq 0 \text{ then } y = x^2; \text{ if } 0 \leq x \leq 3 \text{ then } y = 2x\}$.
79. $Q = \{(x, y) \mid x = 5 \text{ and } y = 2\}$.
80. $Q = \{(x, y) \mid xy + y = 1\}$.
81. $Q = \{(x, y) \mid 2x^2 - 3xy = 2 \text{ and } 0 < x < 1\}$.
82. $Q = \{(x, y) \mid A \text{ is a non-empty set, } x \text{ and } y \text{ are sets, and } A \cup x = A \cup y\}$.
83. $Q = \{(x, y) \mid x \text{ is a circle in a fixed plane and } y \text{ is its center}\}$.
84. $Q = \{(x, y) \mid x \text{ is a set and } y = \langle x \rangle\}$.
85. $Q = \{(x, y) \mid x \text{ and } y \text{ are sets and } x \cup y = x\}$.
86. $Q = \{(x, y) \mid x \text{ and } y \text{ are sets, } \varnothing \in x \text{ and } y \cup \langle \varnothing \rangle = x\}$.
87. $Q = \{(u, y) \mid A \text{ is a non-empty set, } u \text{ and } y \text{ are sets and } y = A \times u\}$.

88. $Q = \{(x, y) \mid x$ and y are sets and $x \subset y\}$.
89. $Q = \{(x, y) \mid x$ and y are real numbers and $y = x^4\}$.
90. Given $Q = \{(x, y) \mid x$ is a set and $y = x \cup \langle x \rangle\}$. Assuming that "$u \in v$ and $v \in u$" is a false statement for all u and all v, show that Q and Q^{-1} are functions.
91. Prove that $\varnothing$ is a function with $\mathscr{D}(\varnothing) = \mathscr{R}(\varnothing) = \varnothing$.
92. Show that if A and B are sets, then $A \times B = \varnothing$ if and only if $A = \varnothing$ or $B = \varnothing$ (this includes the possibility that both are empty).
93. Given $Q = \{(u, v) \mid u$ is an ordered pair of sets and v is their intersection$\}$, show whether or not Q and Q^{-1} are functions.

§§ 1.37 to 1.50

94. Let f be that function for which $f(x) = x^2$ whenever $0 \leq x$. Prove that f^{-1} is a function. What is the conventional way of denoting $f^{-1}(x)$? (*Hint:* Compute $f(u) - f(v)$ for $0 \leq u, 0 \leq v, u \neq v$, factor the resulting expression and draw conclusions accordingly.)
95. Let f be so defined that $f(x) = x^{2n}$ for each $x \geq 0$, where n is a fixed positive integer. Prove that f^{-1} is a function.
96. Suppose $f(x) = x^3$ for each real number x. Show that f^{-1} is a function. What is the conventional notation for $f^{-1}(x)$? (*Hint:* Proceed as in Problem 94 above and show that the quadratic factor has no real roots.)
97. Let f be so defined on the set of all real numbers that $f(x) = -x^2$ if $x < 0$ and $f(x) = x^2$ if $x \geq 0$. Prove that f^{-1} is a function.
98. Suppose that f is a function defined on the set of all real numbers with

$$f(x) = x^3 \text{ if } x \text{ is rational};$$

$$f(x) = 2x - 1 \text{ if } x \text{ is irrational}.$$

Prove that f^{-1} is a function. (*Hint:* A rational number is zero or a number of the form p/q, where p and q are positive or negative integers. All other real numbers are irrational. You may assume that sums and products of rationals are again rational; the sum or the product of a rational and an irrational number is irrational.)

99. Let

$$f = \{(x, y) \mid x \text{ is a circle of unit radius in a fixed plane}; \ y \text{ is its center}\}.$$

Prove that f^{-1} is a function.

100. Let

$$f(x) = \frac{x}{x^2 + 1} \text{ for each real number } x,$$

$$g(x) = \frac{1}{x} \text{ for each real number } x \neq 0.$$

(a) Prove that $(f \circ g)(x) = f(x)$ whenever x is real and $x \neq 0$
(b) Compute $(f \circ f)(x)$ for an arbitrary real number x. Reduce your fractions as far as possible.

101. Given $f = \{z \mid z = (1, 3)$ or $z = (5, 2)$ or $z = (6, 1)\}$ and $g = \{z \mid z = (3, 0)$, or $z = (2, -2)$, or $z = (1, 1)\}$, write an expression for $g \circ f$ and another for $(g \circ f)^{-1}$.

102. Given $f = \{(x, y) \mid x \text{ is a set and } y = \langle x \rangle\}$, prove whether or not $f \circ f$ and $(f \circ f)^{-1}$ are functions.

103. Let $g = \{(x, y) \mid x \text{ is a set and } y = x \cup \langle x \rangle\}$. Prove whether or not $f \circ g$ and $g \circ f$ are functions, where f is defined as in Problem 102 above. Calculate $(f \circ g)(x)$ and $(g \circ f)(x)$ for an arbitrary set x. Are $f \circ g$ and $g \circ f$ the same?

104. With g defined as in Problem 103 above, calculate $(g \circ g)(x)$ for an arbitrary set x.

105. If $f, g,$ and h are functions, $\mathscr{R}(f) \subset \mathscr{D}(g)$, and $\mathscr{R}(g) \subset \mathscr{D}(h)$, then $h \circ (g \circ f) = (h \circ g) \circ f$.

The Real Number System

The concept of *number* is founded on the primitive notion of counting. It is possible to formalize certain basic principles derived from the process of counting, translate these into set-theoretic form, add them to the axioms of set theory, and then, in a series of steps, construct the set known as the real number system. Using ordinary logic, the properties of the real numbers may be derived. In this manner the nature of the real number system is fully exposed, and it is possible to see just how the fundamental properties of the real numbers are obtained. It is then clearly apparent how the logical consistency of the real number system depends on that of the basic axiom system. This kind of program affords the ideal way to develop the real numbers. However, it is extremely time-consuming to carry it out in full. Furthermore, certain aspects of the process are more suitably treated in a course in modern algebra than in one such as this. Consequently, we will not construct the real number system, but we will instead assume its *a priori* existence. We will list a number of its fundamental properties without proof. These may be regarded as axioms, although no special effort will be made to keep their number to a minimum, and some may be derivable from others. Our reason for this is that the effort required to prove some of these properties from the others is much greater than the additional understanding of the system one might gain thereby. For the same reason, we will not stop to prove many of the elementary rules of calculation encountered in algebra; we will assume that they have already been proved. We will, however, discuss and prove results that are not strictly elementary, or that bring out particularly important features of the real number system. The axioms of the real numbers may be given in several rather natural groups, somewhat independent of each other, and they will be so listed.

We shall henceforth assume that we have before us a set R, called the real number system, subject to the properties to be enumerated in §§ 2.1 through 2.6.

2.1 Addition and Multiplication and Their Fundamental Properties. Perhaps the first thing one thinks of in connection with numbers is the idea of addition. We shall assume that addition is given to us along with

R itself. We assume therefore, the existence of a function $+$ called *addition*, whose domain is the set $R \times R$. For an arbitrary ordered pair of real numbers $(x, y) \in R \times R$, the value of this function is $+((x, y))$. However, we shall use the customary notation $x + y$ instead of $+((x, y))$ to denote the value of this function, although the latter, and not the former, is in accord with correct functional notation. With this convention, we list the fundamental properties of addition.[1]

(i) $\mathscr{R}(+) = R$; thus $x + y \in R$ whenever $x \in R$ and $y \in R$ (closure property).

(ii) $x + (y + z) = (x + y) + z$ whenever $x \in R$, $y \in R$, and $z \in R$ (associative property).

(iii) $x + y = y + x$ whenever $x \in R$, and $y \in R$ (commutative property).

(iv) There exists a real number 0, such that $0 + x = x$ whenever $x \in R$. We note that if $0' \in R$ and if $0'$ satisfies

(iv') $0' + x = x$ whenever $x \in R$,

then $0 = 0'$. For, setting $x = 0'$ in (iv) and $x = 0$ in (iv') we obtain, respectively, $0 + 0' = 0'$ and $0' + 0 = 0$. From (iii) we conclude that $0 = 0'$. This justifies giving 0 the special name *zero*. Also it leads to the next axiom.

(v) For each $x \in R$, there exists $y \in R$ such that $x + y = 0$. We let

$$N = \{(x, y) \mid x \in R, y \in R, \text{ and } x + y = 0\},$$

and suppose that $(x, y) \in N, (x, z) \in N$. Using (iv), (ii), and (iii) we see that

$$y = 0 + y = (x + z) + y = (z + x) + y = z + (x + y) = z + 0$$

$$= 0 + z = z.$$

Consequently, N is a function. For any $x \in R$, there exists y such that $(x, y) \in N$ by (v), and $y = N(x)$. For each such x, $N(x)$ is called the *negative of x* and it is customary to denote it by $-x$. Thus, for any $x \in R$, $-x$ is that real number for which $x + (-x) = 0$. Since we have already used $-$ for set subtraction, this is a violation of proper use of notation. However, it causes no confusion, and the custom is of long standing, so we will use $-$ in this new sense anyway.

Associated with addition is the function called *subtraction*, which we momentarily denote by S. By definition, S is the function whose domain is $R \times R$, such that if $x \in R$ and $y \in R$, then $S(x, y) = x + (-y)$. It is

---

[1] We shall observe another convention henceforth. If f is any function whose domain is a set of ordered pairs, and if (x, y) is an ordered pair in its domain, then we will write the value of f at (x, y) as $f(x, y)$ instead of the correct but more cumbersome $f((x, y))$.

customary to drop the symbol $+$ and so to write $S(x, y) = x - y = x + (-y)$. Although this compounds our violation of proper usage of notation, it leads to no misunderstandings, and we will follow the practice.

Any mathematical system satisfying the axioms (i) to (v) is called a *commutative group*. Such systems have been studied extensively by mathematicians. Since R is an example of such a system, it enjoys all properties common to general commutative groups. However, R has many additional properties, as we shall see. The above axioms tell us nothing about multiplication, to which we now turn. We assume the existence of a function called multiplication, whose domain is the set $R \times R$, subject to the axioms (vi) to (xi) below. For any ordered pair $(x, y) \in R \times R$, the value of this function is $\cdot (x, y)$. However, we shall agree to denote it by the conventional $x \cdot y$, or simply xy. We now list the fundamental properties of multiplication.

(vi) $\mathscr{R}(\cdot) = R$; thus $x \cdot y \in R$ whenever $x \in R$ and $y \in R$ (closure property).

(vii) $x \cdot (y \cdot z) = (x \cdot y) \cdot z$ whenever $x \in R$, $y \in R$, and $z \in R$ (associative property).

(viii) $x \cdot y = y \cdot x$ whenever $x \in R$ and $y \in R$ (commutative property).

(ix) There exists an element $1 \in R$ such that $0 \neq 1$ and $1 \cdot x = x$ whenever $x \in R$.

We note that if $1' \in R$ and if $1'$ satisfies

(ix') $1' \cdot x = x$ whenever $x \in R$,

then $1 = 1'$. To show this, we set $x = 1'$ in (ix) and $x = 1$ in (ix'), and obtain, respectively, $1 \cdot 1' = 1'$ and $1' \cdot 1 = 1$. We apply (viii) to conclude that $1 = 1'$.[2] We give 1 the special name *one*. It is sometimes called the *unit* element of R. The present observations give point to the next axiom.

(x) If $0 \neq x \in R$ then there exists a real number y for which $x \cdot y = 1$.

We let

$$J = \{(x, y) \mid 0 \neq x \in R \text{ and } x \cdot y = 1\},$$

and suppose that $(x, y) \in J$, $(x, z) \in J$. Then, using (ix), (vii), and (viii), we see that

$$y = 1 \cdot y = (x \cdot z) \cdot y = (z \cdot x) \cdot y = z \cdot (x \cdot y) = z \cdot 1 = 1 \cdot z = z.$$

Thus J is a function. If $x \in R$, then by (x) there exists y such that $(x, y) \in J$, and therefore $y = J(x)$. Thus if $0 \neq x \in R$, then $J(x)$ is a real number called the *reciprocal* of x. It is customary to denote it by $1/x$ or x^{-1}. If $0 \neq x \in R$, then $1/x$ is that real number for which $x \cdot (1/x) = 1$.

Associated with multiplication is the function of *division*, which we denote temporarily by D. By definition, D is the function defined on $R \times (R - \langle 0 \rangle)$ in such a way that $D(x, y) = x \cdot (1/y)'$ whenever $x \in R$,

---

[2] This is the same technique we used in proving that zero is unique.

$0 \neq y \in R$. We agree to shorten this expression by writing $D(x, y) = x/y = x \cdot (1/y)$.

It may be noted that the rules governing multiplication differ basically from those governing addition only in the fact that one real number, namely 0, has no multiplicative inverse (i.e., reciprocal), whereas each real number has an additive inverse (i.e., negative).

So far there has been no axiom connecting addition and multiplication. This is accomplished as follows.

(xi) $x \cdot (y + z) = x \cdot y + x \cdot z$ whenever $x \in R, y \in R$, and $z \in R$

(distributive property of multiplication with respect to addition).

This completes the list of axioms of ordinary addition and multiplication. From these axioms it is possible to derive all the familiar rules learned in elementary algebra for the removal of parentheses, multiplication and factorization of algebraic expressions, reduction of fractions, and the like.

We will not stop to verify these things, but we will assume that they have already been established, and in the sequel we will feel free to avail ourselves of all the technique of computation normally developed in elementary algebra courses. The sole exception to this will be found in the exercises at the end of this chapter, where the student may acquire a little practice proving some elementary properties using only the axioms and properties otherwise established in this section.

Any mathematical system satisfying the basic axioms (i) to (xi) inclusive is called a *field*. R is evidently an example of a field.

We conclude this section with a final observation about addition and multiplication. The reader may have noticed that although we have listed a number of properties of these functions, *nowhere did we state what is the value of either $x + y$ or $x \cdot y$*, where x and y are arbitrary members of R. It should be clear that simply writing $x + y$ and $x \cdot y$ for the sum and product, respectively, of x and y does not tell us what the sum and product are. This omission may seem odd, since it was stated in the opening chapter that for sets in general, and thus for functions in particular, there must be a criterion which determines just which objects belong to them and which do not. In the case of a function this may take the alternative form of some expression that determines the value of the function associated with an arbitrary member of its domain. Since $+$ and $\cdot$ are functions defined on $R \times R$, there must exist expressions that tell us the values of $x + y$ and $x \cdot y$, respectively, for an arbitrary ordered pair (x, y) belonging to $R \times R$. One may well wonder why such expressions were not given, and why the axioms (i) to (xi) were not proved as consequences of them.

One reason for this is the fact that addition and multiplication, which are defined first on the set of all ordered pairs of positive integers and are later extended to the set of all ordered pairs of real numbers, are usually defined initially by a process known as *definition by induction*,[3] which does not yield explicit formulas for $x + y$ and $x \cdot y$, where x and y are arbitrary positive integers. As these functions are extended step by step, the implicit character of the initial definitions is carried into the extensions, and so explicit expressions for $x + y$ and $x \cdot y$ are never given. Even when addition and multiplication are defined without the use of definition by induction, the process is so involved it would not be easy to write expressions for $x + y$ and $x \cdot y$, to say nothing of verifying the properties (i) through (xi).

2.2   The Order Relation in the Real Number System.   Everyone is familiar with the fact that we rank the real numbers according to their "size." Even the negative numbers are so ranked. A pictorial means of determining the relative ranking of any pair of different real numbers can be obtained by putting all the real numbers on a line, as we do in setting up a coordinate system. In this arrangement, of two given numbers, the one that is farther to the left is the "smaller" of the two. However, we need a more rigorous foundation for the order relation than this, and we now accomplish this by stating a few of its basic properties.

Accordingly, we assume the existence of an *inequality* or *order* relation that is a subset of $R \times R$. If $(x, y) \in R \times R$, we agree to write $x < y$ (read: x *is less than* y) whenever $(x, y) \in \, <$, and $x \not< y$ whenever $(x, y) \notin \, <$. This is in conformity with long-established custom. With this convention established, we now list the axioms of order.

(i) If $x \in R$ and $y \in R$, then exactly one of the following statements is true: $x = y$, $x < y$, $y < x$ (trichotomy property).

(ii) If $x \in R$, $y \in R$, $z \in R$, $x < y$, and $y < z$, then $x < z$ (transitive property).

(iii) If $x \in R$, $y \in R$, $z \in R$, and $x < y$, then $x + z < y + z$.

(iv) If $x \in R$, $y \in R$, $z \in R$, $x < y$, and $0 < z$, then $xz < yz$.

We note that (iii) and (iv) tie the order relation to addition and multiplication. In (iv), we observe that z has to satisfy a special condition. For any real number z, if $0 < z$, we say that z is *positive*, if $z < 0$, we say that z is *negative*. From (i), it follows that every non-zero real number is either positive or negative.

We find it convenient to introduce a somewhat weakened order relation, based on the one already introduced. For this purpose, we agree that $x \leq y$ (read: x *is less than or equal to* y) if and only if $x < y$ or $x = y$.

---

[3] This subject is studied in Chapter 6.

Many mathematicians use the expressions "x is less than y" and "y is greater than x" synonymously. There is a corresponding notation to go with the latter expression. We thus agree that $y > x$ (read: *y is greater than x*) if and only if $x < y$. Similarly, $y \geq x$ (read: *y is greater than or equal to x*) if and only if $x \leq y$.

We mention another convention. Expressions such as "$x < y < z$" are used as abbreviations for "$x < y$ and $y < z$". Similar meanings are to be attributed to corresponding expressions in which $<$ is replaced at one or both places by $\leq$.

We now prove a few useful properties of inequalities from the above axioms. In what follows, we assume that x, y, z, and w are arbitrary real numbers.

(v) If $x < y$ and $w < z$, then $x + w < y + z$. For we may use (iii) twice to obtain $x + w < y + w$ and $w + y < z + y$. Applying (ii), we obtain $x + w < y + z$.

(vi) If x and y are positive, then so is $x + y$. For $0 < x$ and $0 < y$, so by (v), $0 = 0 + 0 < x + y$; hence $x + y$ is positive.

(vii) If $0 < x < y$, $0 < w < z$, then $xw < yz$. For by (ii), since $0 < x$ and $x < y$, then $0 < y$. Applying (iv) twice, we infer that $xw < yw$ and $wy < zy$. From (ii), we conclude that $xw < yz$.

(viii) If x and y are positive, then so is xy. To show this, we see that $0 < y$ and $0 < x$. Hence by (iv), $0 = 0 \cdot x < x \cdot y$. Thus xy is positive.

(ix) If x is positive then $-x$ is negative, and conversely. If x is negative then $-x$ is positive, and conversely. To show this, suppose $0 < x$; then by (iii) $-x = 0 + (-x) < x + (-x) = 0$; $-x$ is negative. If $-x < 0$, then by (iii), $0 = x + (-x) < x + 0 = x$; x is positive. Similarly, if $x < 0$, then by (iii), $0 = x + (-x) < 0 + (-x) = -x$; $-x$ is positive. If $0 < -x$, then by (iii), $x = 0 + x < (-x) + x = 0$; x is negative.

(x) If x and y are negative, then xy is positive. For by (ix), $-x$ and $-y$ are positive, hence by (viii) $(-x) \cdot (-y)$ is positive. But $(-x) \cdot (-y) = xy$ and the proof is complete.[4]

(xi) If x is positive and y is negative, then xy is negative. For by (ix), $-y$ is positive, and so by (viii), $x \cdot (-y)$ is positive. But $x \cdot (-y) = -(xy)$; hence by (ix), xy is negative.

(xii) If $0 \neq x$ then $0 < x^2$.[5] This is an immediate consequence of (viii), (x) and the fact that x must be positive or negative.

(xiii) $0 < 1$. Since $0 \neq 1$, then $0 < 1^2 = 1$ by (xii).

---

[4] We assume that the rules for products of signed numbers are known. See Exercise 8 at the end of this chapter.

[5] We are using the abbreviation $x^2 = x \cdot x$. Similarly, we shall use the exponential notation to denote positive integral powers of mathematical expressions without further explanation in the future.

(xiv) If $0 < x$ then $0 < (1/x)$. For $x \cdot (1/x) = 1$, and 1 is positive. Thus by (xi) and the fact that $x \cdot 0 = 0 \neq 1$ we conclude that $1/x$ must be positive.

(xv) If $0 < x < y$ then $0 < (1/y) < (1/x)$. Since x and y are positive, so is xy, and thus so is $(1/xy)$. Using (iv) we obtain $x \cdot (1/xy) < y \cdot (1/xy)$. Reducing by the usual methods, we obtain $(1/x) < (1/y)$. We infer that $0 < (1/x)$ by (xiv), thus completing the proof.

In some of the foregoing statements, it is possible to replace $<$ by $\leq$ at certain places in the hypotheses, and obtain corresponding results with $\leq$ in the conclusions; sometimes even the strict inequality $<$ may hold in the conclusions, depending upon exactly where the replacement of $<$ by $\leq$ occurred in the hypotheses. The possible combinations are too numerous to consider and list separately; consequently, whenever one of these minor variants is needed in the proof of a theorem, reference (if any is given) will be made to the appropriate statement above, even though it may not be strictly applicable. It will be left to the reader to make the suitable modification in the reference, and to confirm the validity of the application.

We complete the above list with a result of great importance.

(xvi) If $0 \leq x$ and $x \leq \epsilon$ for each $\epsilon > 0$, then $x = 0$. We suppose $x \neq 0$; then we must have $0 < x$. It follows from (xiii), (vi), (xv), and (viii) above that $0 < (x/2)$.[6] Consequently, if we take the particular value $\epsilon = x/2$, we must have $0 < x \leq x/2$, whence by (iv) and (iii), we obtain $0 < 2x \leq 2(x/2) = x$; $2x + (-x) \leq (x - x) = 0$ and so $x \leq 0$. But this contradicts our assumption that $0 < x$. Hence $x = 0$ as required.

It is worth noting that if $x \leq \epsilon$ is replaced by $x < \epsilon$ in the hypotheses of (xvi), then the conclusion is still valid, since this substitution merely strengthens the hypotheses. It is in this revised form that most applications will be made later.

2.3 The Absolute Value Function.

We will now introduce a most useful function whose domain is the set of all real numbers. It is conventional to denote the value of this function for an arbitrary real number x by $|x|$, called the *absolute value of* x. The absolute value function is defined by

$$|x| = x \text{ if } 0 \leq x; \qquad |x| = -x \text{ if } x < 0.$$

A few properties of this function are now given.

(i) If $x \in R$ then $0 \leq |x|$; $x = 0$ if and only if $x = 0$. This is an immediate consequence of the properties of the order relation in § 2.2.

(ii) If $x \in R$ and $y \in R$, then $|xy| = |x| \cdot |y|$.

---

[6] We tacitly define $2 = 1 + 1$. Similarly we define $3 = 2 + 1$, $4 = 3 + 1$, etc., and use these symbols without further explanation in the sequel.

This may be proved by considering separately the cases in which x and y are both positive, both negative, one is positive and the other is negative, and finally the case wherein at least one of them is zero.

(iii) If $x \in R$, $0 \neq y \in R$, then $|x/y| = |x|/|y|$. To show this, we let $z = x/y$. Clearly $z \cdot y = x$; and by (ii), we obtain $|x| = |z \cdot y| = |z| \cdot |y| = |x/y| \cdot |y|$. Dividing by $|y|$, we obtain the desired result.

(iv) If $x \in R$ then $|x| = |-x|$, $-|x| \leq x \leq |x|$. This is established by considering separately the cases where x is positive, negative, and zero.

(v) If $x \in R$ and $y \in R$, then $|x + y| \leq |x| + |y|$. This is the so-called *triangle inequality* and is of great importance. Its validity is established by considering separately the cases in which x and y are both positive, both negative, one is positive and the other is negative, and finally the case in which at least one of them is zero.

(vi) If $x \in R$ and $y \in R$, then $\big| |x| - |y| \big| \leq |x - y|$. To prove this, we note from (v) and (iv) that $|x| \leq |x - y| + |y|$; hence $|x| - |y| \leq |x - y|$. Similarly, $|y| \leq |y - x| + |x| = |x - y| + |x|$; hence $-(|x| - |y|) \leq |x - y|$. Now by definition of the absolute value, one of the two quantities $|x| - |y|$ and $-(|x| - |y|)$ is the absolute value of $|x| - |y|$. Thus from above, we must have $\big| |x| - |y| \big| \leq |x - y|$.

(vii) If $x \in R$, $\epsilon \in R$ and $0 < \epsilon$, then the relation $|x| < \epsilon$ is equivalent to the simultaneous inequalities $-\epsilon < x < \epsilon$; similarly $|x| \leq \epsilon$ is equivalent to $-\epsilon \leq x \leq \epsilon$. This is easily established by considering the cases in which x is positive, negative, and zero.

2.4 The Positive Integers and Their Basic Properties.

If one builds up the number system starting with the positive integers and zero, then the usual procedure consists of adding the negative integers to obtain the *whole number* system; then fractions are added to this system to obtain the *rational number* system; finally the *irrational numbers* are adjoined to the rational number system to obtain the full real number system. Also, as one progressively enlarges the number system, the concepts of addition, multiplication, and order are extended to the larger systems in a certain manner. However, since we adopt the point of view that the real number system has already been built up, and we are reciting a partial though basic list of properties possessed by it, we find it necessary to go into this system to find the positive integers that were used in building it. Consequently, we list separately the properties of a special subset of R, namely P, the set of *positive integers*, as follows:

(i) $1 \in P$.

(ii) If $x \in P$, then $x + 1 \in P$.

(iii) If $x \in P$ and $y \in P$, then $x + y \in P$ and $xy \in P$.

(iv) For any x, if $x \in P$, then $1 \leq x$; equivalently there does not exist y such that $y \in P$ and $y < 1$.

(v) If $x \in P$, $y \in P$ and $x < y$, then $(y - x) \in P$.

(vi) If $x \in P$, then there does not exist y such that $y \in P$ and $x < y < x + 1$.

The final fundamental property of P that we state as an axiom is the so-called *well-ordering principle*. This is a most important concept. In order to prepare for its statement, we need a definition. If S is a non-empty set of real numbers, and if there exists in S an element y for which $x \leq y$ whenever $x \in S$, then we say that y *is a largest element of S*. If y and z belong to S and have the property that $x \leq y$ and $x \leq z$ whenever $x \in S$, then clearly $z \leq y$ and $y \leq z$, consequently $y = z$. From this observation it follows that if S has any largest element at all, the set of largest elements of S is a singleton, and consequently we are justified in referring to its only member as *the largest member of S*. Similarly, if $S \neq \varnothing$, and if there exists in S an element y such that $y \leq x$ whenever $x \in S$, then we say that y *is a smallest member of S*. If there exist elements y and z in S such that $y \leq x$ and $z \leq x$ whenever $x \in S$, then clearly $y \leq z$ and $z \leq y$, so that $y = z$. Thus if S has a smallest element, it is unique, and we refer to it as *the smallest member of S*.

It is easy to find sets of real numbers that have neither a smallest nor a largest member; the set of all positive and negative integers is one such set. However, suppose that we consider the set P. Let us take a non-empty subset Q of P. If we imagine that the positive integers are marked in sequence at a uniform distance apart on a horizontal line, increasing as we go to the right, then it seems intuitively apparent that if we traverse this line to the right, we must eventually encounter at some point a "first" element of Q. Such an element will be, as we see intuitively, the smallest element of Q. Thus on intuitive grounds, it appears that every non-empty set Q of positive integers should have a smallest member. Of course, the heuristic argument just given does not prove that every non-empty set Q of positive integers has a smallest member, but it lends plausibility to the idea, which we now formulate as an axiom.

(vii) If $\varnothing \neq Q \subset P$, then Q has a smallest member. This is the *principle of well-ordering of the positive integers*.

Another very important property of the positive integers, sometimes called the *axiom of induction*, will now be added to the list. However, in this case we will state and prove this property from those already listed, making essential use of (vii).

(viii) If $Q \subset P$, $1 \in Q$, and $x + 1 \in Q$ whenever $x \in Q$, then $Q = P$. To prove this, we suppose, if possible, that $Q \neq P$. We let $T = P - Q$, note that $T \neq \varnothing$, use (vii) to infer that T has a smallest member z, and observe that $z \notin Q$. Since $1 \in Q$ by assumption, then $z \neq 1$, and so by (iv), $1 < z$. We set $w = z - 1$ and note from (v) that $w \in P$. Since $w < w + 1 = z$, and z is the smallest member of T, then $w \notin T$; consequently we must have $w \in Q$. But then $z = w + 1 \in Q$, by the hypotheses on Q. This contradicts our earlier observation that $z \notin Q$. The difficulty arises from our initial assumption that $Q \neq P$. Thus $Q = P$, and the proof is complete.

We also state and prove a variation of the axiom of induction that is useful in some situations.

(ix) If $Q \subset P$, and $x \in Q$ whenever $\{t \mid t \in P \text{ and } t < x\} \subset Q,$[7] then $Q = P$.

We suppose, if possible, that $Q \neq P$, let $T = P - Q$, note that $T \neq \varnothing$, and use (vii) to infer that T has a smallest member z. Thus, for any t, if $t \in P$ and $t < z$, then $t \notin T$, whence we must have $t \in Q$. Consequently, we infer that $\{t \mid t \in P \text{ and } t < z\} \subset Q$. Due to the hypotheses on Q, we conclude that $z \in Q$. But $z \in T = P - Q$, so that $z \notin Q$. The contradiction we have just derived forces the conclusion that $Q = P$.

The axiom of induction will be used repeatedly in the development of our subject. It is the basis of the process known as *proof by induction*. This method of proof is particularly well suited to the establishment of theorems that are true for all positive integers. We now give some applications of this method.

For an arbitrary positive integer n, consider the number $n^3 + 2n$. If we replace n by a few specific positive integers, it will be found each time that the resulting expression is divisible by 3. This suggests that $n^3 + 2n$ is divisible by 3 for each positive integer n; that is, there exists k such that either $k = 0$ or $k \in P$, and $n^3 + 2n = 3k$. We will prove this conjecture by induction.

Accordingly, we let

$$Q = \{n \mid n \in P \text{ and } n^3 + 2n \text{ is divisible by } 3\}$$

We first replace n by 1 in the statement defining Q, and note that the resulting statement is true; $1 \in Q$. The second part of a proof by induction always involves showing that for any x, if $x \in Q$, then $x + 1 \in Q$. Accordingly, we assume x is an arbitrary element of Q, and we attempt to prove that necessarily $x + 1$ must be an element of Q.

---

[7] The condition $1 \in Q$ in (viii) is essential; otherwise we could take $Q = \varnothing$ satisfying the remaining hypotheses, and we could not prove $Q = P$. In (ix), it may appear that we have overlooked this point. However, if we replace x by 1 we see that $\{t \mid t \in P \text{ and } t < 1\} \subset Q$ is a true statement, since $\{t \mid t \in P \text{ and } t < 1\} = \varnothing$. Thus $1 \in Q$ due to our hypotheses on Q.

The assumption that $x \in Q$ means that there exists k, $k = 0$ or $k \in P$, for which

(1) $$x^3 + 2x = 3k.$$

To show that $x + 1 \in Q$, we must prove that $(x + 1)^3 + 2(x + 1)$ is divisible by 3. To this end, we compute

(2) $$(x + 1)^3 + 2(x + 1) = x^3 + 3x^2 + 5x + 3$$
$$= (x^3 + 2x) + (3x^2 + 3x + 3)$$
$$= (x^3 + 2x) + 3(x^2 + x + 1).$$

The grouping of the terms in (2) was done with (1) in mind. For, since we are assuming the validity of (1), we can substitute into (2) to obtain

(3) $(x + 1)^3 + 2(x + 1) = 3k + 3(x^2 + x + 1) = 3(k + x^2 + x + 1).$

Since $k = 0$ or $k \in P$, and $x \in P$, it follows readily that $(k + x^2 + x + 1) \in P$; and thus the right-hand side of (3) is divisible by 3. Thus $(x + 1) \in Q$. It should be kept in mind that this last conclusion was based upon the assumption that $x \in Q$. However, this is all that we require to apply (viii) and conclude that $Q = P$. Thus it is seen that $n^3 + 2n$ is divisible by 3 for each $n \in P$, as we set out to prove.

We consider another example. We shall say that a positive integer n is *even* if and only if there exists a positive integer k such that $n = 2k$; we shall say that n is *odd* if and only if there exists a positive integer m such that $n = 2m - 1$. We wish to show that every positive integer is even or odd.

To this end, we let

$$Q = \{n \mid n \in P \text{ and } n \text{ is even or } n \text{ is odd}\}.$$

We replace n by 1 and note that if we let $m = 1$, then $1 = 2m - 1$, whence 1 is odd, and $1 \in Q$. Next, we assume that $x \in Q$, and try to prove that $x + 1$ necessarily belongs to Q as a consequence of this assumption.

The assumption that $x \in Q$ means that x is even or x is odd. If x is even, then there exists $k \in P$ such that $x = 2k$; thus $x + 1 = 2(k + 1) - 1$; we see that $x + 1$ is odd, and so $x + 1 \in Q$. If x is odd, there exists $m \in P$ such that $x = 2m - 1$; thus $x + 1 = 2m$, $x + 1$ is even, and so $(x + 1) \in Q$. Thus, in any case, if $x \in Q$, then $x + 1 \in Q$. Applying (viii), we conclude that $Q = P$, and so every positive integer is even or odd.

We give a final example of a proof by induction. Let us suppose we are given a real number C and the sequence[8] of real numbers X, such that

---

[8] For a definition of *sequence*, see the beginning of Chapter 4.

$X_n < X_{n+1}$ for each positive integer n and $C < X_1$. We shall prove that $C < X_n$ holds for each positive integer n.

Before doing so, however, we wish to discuss the necessity for giving a formal proof by induction of this fact, which may seem obvious enough without such a proof. For one might argue as follows: X_1 is less than X_2 and C is less than X_1, therefore C is less than X_2; but also X_2 is less than X_3, therefore C is less than X_3, and so on. This kind of argument, in which the ultimately desired result is not established, but instead a process of "proof" is merely indicated, may simply mask the real difficulty. For a proper proof of any expression must consist of a finite sequence of statements, each following from its predecessors in accordance with the rules of logic, and with no steps omitted.[9] In the case at hand, by successive repetitions of the procedure suggested above by the phrase "and so on," we could prove $C < X_4$, $C < X_5$, and even $C < X_{1000}$, by writing out all the steps in a manner that would satisfy the most fastidious logician. In fact, we can see intuitively that if we replace n in the expression "$C < X_n$" by any specific positive integer, the resulting statement could be proved impeccably using the suggested procedure. Thus in such a case, the means of proof indicated by "and so on" does not conceal any difficulty but instead is used to save the time and energy of writing out something that everyone is willing to concede could be written out if the occasion demanded it.

However, we really want to prove the truth of the expression "if $n \in P$, then $C < X_n$," from our hypotheses, and for this purpose, the procedure indicated by "and so on" is totally inadequate. For, although it certainly enables us to prove that $C < X_4$, $C < X_5$, $C < X_6$ etc., nowhere in the continuation of this sequence of sentences will be found the expression "$C < X_n$" because n is a variable, and none of the sentences in the sequence involves variable indices. Furthermore, because there are infinitely many sentences in the sequence, it is impossible to write their totality as a conjunction and so express them ultimately in succinct form using the universal quantifier, which would yield the result we seek. This is why we are forced to seek a technique that does not involve "and so on."

We now give a proof by induction. We let

$$Q = \{n \mid n \in P \text{ and } C < X_n\}.$$

We replace n by 1 and note that $C < X_1$ by assumption, making $1 \in Q$. We next assume that $n \in Q$ and try to show that $n + 1$ must consequently belong to Q. But then $C < X_n$, and by assumption, $X_n < X_{n+1}$, hence

[9] Actually, even logicians seldom give complete proofs. For practical reasons, it becomes necessary to abbreviate and omit some steps. However, these abbreviations and omissions are the kind that can easily be filled in.

$C < X_{n+1}$, whence $n + 1 \in Q$. From (viii), we now infer that $Q = P$ and so the desired statement has been proved.

We conclude this section with the proof of another useful property of the positive integers.

(x) If $\varnothing \neq T \subset P$ and if there exists a positive integer y such that $x \leq y$ for each $x \in T$, then T has a largest member.

To show this, we let

$$S = \{u \mid u \in P \text{ and } x \leq u \text{ whenever } x \in T\}.$$

From our hypotheses $S \neq \varnothing$ and so by (vii), S has a smallest member z. Now $x \leq z$ for each $x \in T$. If $z \in T$, then z is evidently the largest member of T and the proof is complete. Accordingly we assume, if possible, that $z \notin T$; then $x < z$ for each $x \in T$. Since $T \neq \varnothing$, there exists $t \in T$, whence $1 \leq t < z$, and so $1 < z$. We let $s = z - 1$; $s \in P$ by (v) above. Since $s < z$ and z is the smallest member of S, then $s \notin S$; consequently there exists an element t' of T such that $s < t'$; also $t' < z = s + 1$. These inequalities are in violation of (vi) above. Thus we conclude that $z \in T$, as we set out to prove.

2.5   The Whole Numbers and the Rationals.   A number x is called a *negative integer* if and only if $-x \in P$. The set of all positive and negative integers and zero is called the *whole number* or *integer* system. A number y is called a *rational number* if and only if there exist whole numbers p and q, $q \neq 0$, such that $y = p/q$. Since every whole number p can be represented in the form $p = p/1$, it follows that the whole number system is a subset of the rational number system.

We do not require any new axioms to express the properties of the whole numbers and the rationals, but we will list below a few properties of these classes of numbers. They are relatively simple consequences of what has gone before. The first requires only a consideration of the signs of the numbers concerned, the second follows from the rules of elementary algebra and application of the first.

(i) The sum, difference, and product of two integers is an integer.

(ii) The sum, difference, and product of two rational numbers is a rational number; so is the quotient if the denominator is not zero.

(iii) Between any two unequal rational numbers there is a rational number different from each.

If $x \in R$, $y \in R$ and $x \neq y$, then it is easy to check that the arithmetic mean of these numbers, namely $(x + y)/2$, lies strictly between x and y. However, if x and y are rational numbers, then from (ii) it follows that their arithmetic mean is again rational. The property (iii) is sometimes expressed by saying that the rational numbers are *dense-in-themselves*. In this respect, the rationals are essentially different from the integers, since between the integers n and $n + 1$ there exists no integer.

2.6 Irrational Numbers and the Completeness Axiom. We have discussed the properties of the positive integers, the whole numbers, and the rationals. The fact that the latter are dense-in-themselves gives one the intuitive feeling that they are very numerous, and that there might not exist any other numbers, or at least that there might exist no reason for trying to enlarge the number system beyond this point. In general, there are at least two reasons for attempting to extend any mathematical system. One is that it sometimes proves to be impossible to formulate or perhaps to solve certain problems of the physical world in a given system. The other is more formal; the mathematician may simply want to see how far extension is possible, preserving as he goes along some properties he regards as important. As it happens, these two aims frequently go hand in hand.

Regarding the number system, at the outset one usually starts with the positive integers or else the *natural numbers*, which are made up of the positive integers together with zero. These numbers are not adequate for the formulation of many real-life situations, and consequently the whole number and rational number systems were devised to enable men to formulate and solve many problems. Mathematicians showed that these extensions of the basic system were logically sound, and they developed the properties of the extended systems. There are relatively simple problems that cannot be solved in the rational number system, so mathematicians have been forced to seek an extension in which these problems can be solved. The resulting number system is called the *real number* system. However, there are still problems of a simple nature that cannot be solved in the real number system, and *complex* numbers were devised for the purpose of answering these and other questions. More complicated number systems called *quaternions* and *octonions* have been invented for special purposes. In the present work, we will not go beyond the real numbers.[10]

We now show with a simple example just why the rational numbers are not adequate for many purposes. We consider a right triangle, each of whose legs is of unit length. Then the hypotenuse is of length $\sqrt{2}$. We shall now see that it is not possible to represent even this simple number in the rational number system. We assume that there exist positive integers p and q such that $p/q = \sqrt{2}$, and we further assume that p and q have no factors in common but 1. Now $p^2/q^2 = 2$, thus $p^2 = 2q^2$. If p were odd,

[10] A slight extension of this system is taken up in Chapter 5.

there would exist a positive integer k such that $p = 2k - 1$, and so $p^2 = 4k^2 - 4k + 1$, whence p^2 would also be an odd integer and not divisible by 2. However, $2q^2$ is divisible by 2. Thus we conclude that p must be even, and so there exists a positive integer m such that $p = 2m$. Then $2q^2 = p^2 = 4m^2$, and so $q^2 = 2m^2$. By a repetition of the argument just given, it follows that q must be even, hence there exists a positive integer n such that $q = 2n$. But this means that p and q have the common factor 2, contrary to hypothesis. Hence it is not possible to represent $\sqrt{2}$ as a rational number of the form p/q in which p and q have no common factors but 1. However, every rational number can be reduced to such a form;[11] therefore $\sqrt{2}$ is not a rational number. Numbers such as this, which belongs to the real number system but are not rational, are called *irrational* numbers.

In similar fashion, it can be shown that such numbers as $\sqrt{3}$, $\sqrt[4]{7}$, $\sqrt[3]{5}$, etc., are irrational. The reason we are able to define these and other irrational numbers is to be found in the *completeness axiom*, which we will state shortly. Before doing so, we will need some definitions.

We take an arbitrary non-empty set S of real numbers. We say that S is *bounded above* if and only if there exists a real number y such that $x \leq y$ whenever $x \in S$. Any number y satisfying this condition is called an *upper bound of S*. Similarly, we say that S is *bounded below* if and only if there exists a real number z such that $z \leq x$ whenever $x \in S$. Any number z satisfying this condition is called a *lower bound of S*. In case S is bounded both above and below, we say simply that S is *bounded*.

Now we let

$$T = \{y \mid y \text{ is an upper bound of } S\},$$
$$V = \{z \mid z \text{ is a lower bound of } S\}.$$

Now $T \neq \emptyset$ if and only if S is bounded above; and $V \neq \emptyset$ if and only if S is bounded below. One may then wonder whether or not T has a smallest member, or whether V has a largest member (recall the definitions in § 2.4).

If T does have a smallest member t, then clearly t is an upper bound of S, and so it is the smallest of all upper bounds of S. For this reason, t will be called the *least upper bound* or *supremum* of S, abbreviated as *l.u.b. S* or *sup S*, respectively. Similarly, if V has a largest member v, then v is a lower bound of S, and it is evidently the biggest of all lower bounds of S. Thus v will be called the *greatest lower bound* or *infimum* of S, abbreviated as *g.l.b. S* or *inf S*, respectively.

---

[11] This is a well-known fact, which we will not prove here, since it does not play a role in the general theory we will develop.

Not all sets of real numbers have an upper bound, consequently such sets have no supremum. Similarly, some sets of real numbers have no infimum. However, the *axiom of completeness*, which we now state, guarantees that some sets of real numbers do have a supremum.

(i) If $\varnothing \neq S \subset R$, and if S is bounded above, then S has a supremum.

This is a very interesting axiom. It does not guarantee that any set of real numbers, even those that are bounded above and below, will necessarily have a largest or a smallest member; but it does say that the set of upper bounds of any given non-empty set of real numbers having at least one upper bound must itself have a smallest member.

Just as the rational number system differs fundamentally from the integers in being dense-in-itself, so the full real number system differs from the rational number system in the property of completeness just expressed.

We can show that the rational numbers do not possess the completeness property by considering the sets

$$S = \{y \mid y \text{ is rational}, 0 \leq y, \text{ and } y^2 < 2\},$$
$$T = \{z \mid z \text{ is rational}, 0 \leq z, \text{ and } 2 < z^2\}.$$

We note first that $0 \in S$, so that $S \neq \varnothing$. Also, suppose that $s \in S$; then $s^2 < 2$ and $2 - s^2$ is a positive rational number. We select a positive rational number[12] h such that $h < 1$ and $h < (2 - s^2)/(2s + 1)$, and let $u = s + h$. Clearly u is rational and $0 < s < u$. Also we easily check that $u^2 = s^2 + 2sh + h^2 < s^2 + 2sh + h < 2$, whence $u \in S$. We have shown that for any given member of S, there always exists a larger member; hence S has no largest member.

We want to show that T is the set of all rational upper bounds of S. First we note that any such upper bound must be non-negative, since the members of S are all non-negative. Let t denote a rational upper bound of S. Then either $t^2 = 2$, $t^2 < 2$, or $2 < t^2$. The first possibility is ruled out since the equation cannot be satisfied by any rational number. If it were possible that $t^2 < 2$, then t would belong to S and, as we just saw, we could find a member of S exceeding t, and so t would not be an upper bound of S. Thus we must have $2 < t^2$, that is, $t \in T$. Thus all upper bounds of S are members of T. We now take arbitrary members $s \in S$ and $t \in T$. We have then $s^2 < 2 < t^2$, $0 < t^2 - s^2 = (t - s)(t + s)$; since s and t are both non-negative, this requires $0 < t - s$, or $s < t$. Hence t is an upper bound of S. We have thus shown that T coincides with the set of upper bounds of S.

We complete our demonstration by showing that T has no smallest member. For consider an arbitrary element $t \in T$. Then $2 < t^2$ and $t^2 - 2$

---

[12] If x and x' are any positive rational numbers, then $xx'/(x + x')$ is positive, rational, and is less than both x and x'. See Problem 59 of the Exercises at the end of this chapter.

is a positive rational number. We let $z = t - (t^2 - 2)/2t = (t/2) + (1/t)$. Clearly z is rational and $0 < z < t$. Also $z^2 = (t^2/4) + 1 + (1/t^2)$. By differential calculus, one can show easily that $2 \leq z^2$. The equality is ruled out since z is rational, hence $2 < z^2$, and so $z \in T$. This shows that S has no rational least upper bound.

Let us now modify the definitions of S and T by deleting the requirement that their members must be rational; we let S' and T', respectively, be the corresponding sets. Now, by dropping the references to rationals wherever these occur, the proofs already given can be used to show that

(1) S' has no largest member and consequently no member of S' is an upper bound of S'.

(2) all members of T' are upper bounds of S'.

(3) T' has no smallest member.

But now, since we are dealing with real and not merely rational numbers, S' has a non-negative least upper bound k. By (1), $k \notin S'$, hence $2 \leq k^2$. If $k \in T'$, we could use (3) to find an upper bound of S' that is smaller than k, which is the least upper bound. Thus $k \notin T'$, whence $k^2 \leq 2$ and we are led to conclude that $k^2 = 2$. By the very definition of square roots, this means that $k = \sqrt{2}$. We have thus shown that $\sqrt{2}$ is a real number. In similar fashion, we could establish that $\sqrt{3}, \sqrt[4]{7}, \sqrt[3]{5}$, etc., are real numbers. As it happens, there exist better ways of proving that such entities are real numbers, from a consideration of inverses of certain continuous functions (see Chapter 7). There are many other real numbers that are not solutions of simple algebraic equations; such numbers are called *transcendental* real numbers.

We devote the rest of this chapter to the proof of some general theorems on real numbers.

2.7 Theorem. If x and y are positive real numbers, then there exists a positive integer N such that $y < Nx$ (*Archimedean property*).

Proof. We prove this by contradiction, and accordingly we assume the contrary; namely, that $nx \leq y$ for each positive integer n. Thus the non-empty set

$$S = \{u \mid \text{there exists a positive integer } n \text{ such that } u = nx\}$$

is bounded above by y; and so it has a least upper bound z. Thus $nx \leq z$ holds for each $n \in P$; consequently $(n + 1)x \leq z$ holds for each $n \in P$, since $n + 1 \in P$ whenever $n \in P$. Thus $nx + x \leq z$ and so $nx \leq z - x < z$ holds for each $n \in P$. However, this means that $z - x$ is an upper bound for S, in contradiction to the fact that z is the least upper bound of S.

2.8 Corollary. If y is any real number, there exists $N \in P$ for which $y < N$.

Proof. If $y \leq 0$, we may take $N = 1$. If $0 < y$, the desired result follows as a special case of Theorem 2.7 with $x = 1$.

2.9 Corollary. If $x \in R$, then there exist integers M and N for which $M < x < N$.

Proof. If $0 \leq x$, we may take $M = -1$ and determine the existence of N in accordance with Corollary 2.8. If $x < 0$, then $0 < -x$ and we apply what we just learned to infer the existence of integers M' and N' for which $M' < -x < N'$; then $-N' < x < -M'$. Clearly, if we put $M = -N'$ and $N = -M'$, then M and N satisfy the stated conditions.

2.10 Corollary. If x is any real number, then there exists a unique integer N such that $N \leq x < N + 1$.

Proof. We first show the existence of the required integer by an examination of three separate cases, and then prove its uniqueness.

CASE 1. $0 \leq x < 1$. Here we take $N = 0$.

CASE 2. $1 \leq x$. We let

$$S = \{n \mid n \in P \text{ and } n \leq x\}.$$

Since $1 \in S$ then $S \neq \varnothing$. Also by Corollary 2.9, there exists an integer N', necessarily positive, such that $x < N'$. But then N' is an upper bound for S, so by (x) of § 2.4, S has a largest member N; thus $N \leq x$. Since $N + 1 \notin S$, then $x < N + 1$ and the proof for Case 2 is complete.

CASE 3. $x < 0$. Then $0 < -x$. From Cases 1 and 2 we see that we can determine the existence of an integer N'' such that $N'' \leq -x < N'' + 1$; hence $-(N'' + 1) < x \leq -N''$. Now if $x = -N''$ we take $N = -N''$; if $x < -N''$ we take $N = -(N'' + 1)$. It is clear that this establishes the proof for Case 3.

We complete the proof by showing that N is unique. For suppose integers N and M satisfy the inequalities

$$N \leq x < N + 1 \text{ and } M \leq x < M + 1.$$

Then

$$N \leq x < N + 1 \text{ and } -(M + 1) < -x \leq -M.$$

Adding these inequalities in accordance with (v) of § 2.2, we obtain

$$(1) \qquad (N - M) - 1 < 0 < (N - M) + 1.$$

Clearly $N - M$ is an integer. If $0 < N - M$ then $(N - M) \in P$ and from the left half of (1) we obtain $0 < N - M < 1$, contrary to (iv) of § 2.4. If $0 < M - N$ then $M - N \in P$, and from the right half of (1) we obtain $0 < M - N < 1$, again a contradiction of (iv) of § 2.4. Thus we infer that $N = M$.

2.11   Corollary.   If ϵ is a positive real number, then there exists a positive integer N such that $(1/N) < \epsilon$.
Proof.   Take $y = 1$, $x = \epsilon$, and apply Theorem 2.7.

In § 2.5 it was shown that the rational numbers are dense-in-themselves; that is, between any two unequal rational numbers, there exists a rational number different from both. We now show that the rationals are dense in the reals; that is, between any two unequal real numbers there exists a rational number.

2.12   Theorem.   If $x \in R$, $y \in R$, and $x < y$, then there exists a rational number r such that $x < r < y$.
Proof.   We apply Corollary 2.11 with $0 < y - x = \epsilon$ to determine the existence of $q \in P$ such that $0 < 1/q < y - x$. Due to Corollary 2.10 there exists an integer p such that $p \le xq + 1 < p + 1$. This yields

$$\frac{p}{q} \le x + \frac{1}{q} < \frac{p}{q} + \frac{1}{q}.$$

Subtracting $1/q$ from each term in this last relation, and putting together our inequalities, we obtain

$$\frac{p-1}{q} \le x < \frac{p}{q} \le x + \frac{1}{q} < x + (y - x) = y.$$

Evidently p/q is a rational number fulfilling the requirements of the theorem.

We now prove a theorem that is an immediate consequence of the definitions of least upper bound and greatest lower bound, but it is so frequently used it is convenient to formulate it as a theorem for future reference purposes. In this theorem, we employ for the first time a device we will use in the future when it happens to be convenient; namely, we give the proof of two theorems at one time. This will be done only in those cases where the two theorems and their proofs are obtained from each other by minor parallel changes in the wording of each. The proof of one of the theorems is obtained by ignoring the parenthetical expressions; the proof of the other is obtained by substituting the parenthetical expressions at the appropriate places in a rather obvious manner.

2.13   Theorem.   If S is a non-empty set of real numbers bounded above (below), with M as its l.u.b. (m as its g.l.b.), and if $0 < \epsilon$, then there exists $y \in S$ for which $M - \epsilon < y \le M$ ($m \le x < m + \epsilon$).
Proof.   Suppose, if possible, that the theorem were false. Then there would exist a positive number ϵ', such that there exists no element $y \in S$

for which $M - \epsilon' < y \leq M$ $(m \leq y < m + \epsilon')$. Then for each element $x \in S$ we would have $x \leq M - \epsilon'$ $(m + \epsilon' \leq x)$. This would mean that $M - \epsilon'$ $(m + \epsilon')$ is an upper (lower) bound for S, contradicting the fact that M (m) is the l.u.b. (g.l.b.) of S. This proves the theorem.

We now prove a theorem that parallels the axiom of completeness.

2.14 Theorem. If S is any non-empty set of real numbers bounded below, then S has a greatest lower bound.

Proof.   We let T denote the set of all lower bounds of S. By hypothesis, $T \neq \varnothing$. Now $S \neq \varnothing$, so there exists an element $y \in S$; and for any $z \in T$, we have necessarily $z \leq y$. But then T is bounded above, and by the axiom of completeness, T has a least upper bound m. We assert that m is a lower bound for S. For if not, then there would exist $x \in S$ such that $x < m$; taking $\epsilon = m - x$, we use Theorem 2.13 to infer the existence of $z \in T$ such that $x = m - \epsilon < z \leq m$, in contradiction to the fact that z is a lower bound of S. Thus m is a lower bound of S, and if m' is any other lower bound then $m' \in T$, and $m' \leq m$ since m is the least upper bound of T. Hence m is the greatest lower bound of S.

Exercises

§ 2.1

1. Show from the definition alone that if $x \in R$, then $-(-x) = x$.
2. From the definition alone, show that $-0 = 0$.
3. Show that if x, y, and z belong to R and if $x + y = x + z$, then $y = z$. (*Hint:* Add $-x$ to both sides of the given equation and reduce the resulting expressions using only the axioms.)
4. Prove from the definition alone that if $0 \neq x \in R$, then the reciprocal of $1/x$ is x.
5. Prove that the reciprocal of 1 is 1, from the definition.
6. Show that if $0 \neq x \in R$, $y \in R$, $z \in R$, and $xy = xz$, then $y = z$. (*Hint:* Multiply both sides of the equation by $1/x$ and reduce the resulting expression using the axioms only.)
7. Prove that if $x \in R$ and $y \in R$ then $-(x \cdot y) = (-x) \cdot y$. (*Hint:* Note that $x + (-x) = 0$. Multiply both sides of this equation by $-y$ and draw conclusions accordingly.)
8. Show that if $x \in R$ and $y \in R$, then $(-x) \cdot (-y) = x \cdot y$.
9. Prove that $0 \cdot x = 0$ whenever $x \in R$. (*Hint:* $0 \cdot x = (0 + 0) \cdot x = 0 \cdot x + 0 \cdot x$. Proceed from this point.)
10. Show that $(-1) \cdot x = -x$ whenever $x \in R$. (*Hint:* $1 + (-1) = 0$. Multiply both sides of this equation by x and draw conclusions accordingly.)

§ 2.2

11. Show that if x and y are negative real numbers, then $x + y$ is negative.
12. Show that if x and y are real numbers and $x < y$, then $-y < -x$.
13. Show that if x, y, w, and z are real numbers, $x \leq y$ and $w < z$, then $x + w < y + z$. Prove that if we know only that $w \leq z$, then we can conclude only that $x + w \leq y + z$.

14. Prove that if x, y, w, and z are real numbers, $0 < x \leq y$ and $0 < w < z$, then $wx < yz$. Show that if we know only that $0 < w \leq z$, then we can conclude only that $wx \leq yz$.

15. Prove that if $0 < x \leq y$ then $(1/y) \leq (1/x)$.

16. Show that if x and y are such real numbers that $x \leq y$ and $y \leq x$, then $x = y$.

17. Show that if x is a real number, then $x^2 = 0$ if and only if $x = 0$.

18. If x, y, w, and z are real numbers, $0 < x < y$ and $0 < w < z$, then $(x/z) < (y/w)$.

19. If x, y, and z are such real numbers that $x < y$ and $z < 0$, then $yz < xz$.

20. Prove that if x is any negative real number, then $1/x$ is negative.

21. Let x and y be such real numbers that $x < y$. Show that their arithmetic mean $z = (x + y)/2$ satisfies the relation $x < z < y$.

22. Solve for all real values of x that satisfy the inequality $x^2 - 5x > 6$, that is, find one or more relations in x alone that are logically equivalent to the given one. (*Hint:* Consider the equivalent inequality $x^2 - 5x - 6 > 0$ that factors into another equivalent relation $(x - 6) \cdot (x + 1) > 0$. Now note that for this inequality to hold, both factors must be positive, or both negative. Use these facts to infer finally that the desired solution is $6 < x$ or $x < -1$. It may be noted that in finding all solutions of the given inequality, we are effectively proving the equality of two sets; namely $\{x \mid x \text{ is real and } x^2 - 5x > 6\} = \{x \mid x < -1 \text{ or } 6 < x\}$.)

23. Solve for all real values of x satisfying the relation $x^2 + 4x < 5$.

24. Solve for all real values of x satisfying the inequality $x^3 - 3x^2 - 4x < 0$.

25. Solve for all real values of x satisfying the inequality $(x - 1)(x + 2)(x - 5) < 0$.

26. Solve for all real values of x satisfying the relation $(x - 3)/(x - 2) < 2$. (*Hint:* The desired inequality does not hold if $x = 2$. Thus we may assume $x \neq 2$ and multiply both sides of the resulting inequality by the positive number $(x - 2)^2$ without changing the direction of the inequality. The resulting expression can be handled as in the preceding problems.)

§ 2.3

27. Carry out the verification of (i) in § 2.3.

28. Give the proof of (ii) in § 2.3.

29. Prove (iv) in § 2.3.

30. Prove (v) in § 2.3.

31. Prove (vii) in § 2.3.

32. If x and ϵ are real numbers, show that the inequality $0 < \epsilon < |x|$ implies $\epsilon < x$ or $x < -\epsilon$, and conversely. Similarly, show that $0 < \epsilon \leq |x|$ implies $\epsilon \leq x$ or $x \leq -\epsilon$, and conversely.

33. Show that if x, y, and z are real numbers, then $|x + y + z| \leq |x| + |y| + |z|$.

34. Given that x, c, and ϵ are real numbers and $0 < \epsilon$, show that the relation $|x - c| < \epsilon$ is equivalent to $c - \epsilon < x < c + \epsilon$, $|x - c| \leq \epsilon$ is equivalent to $c - \epsilon \leq x \leq c + \epsilon$; $|x - c| > \epsilon$ is equivalent to $x < c - \epsilon$ or $c + \epsilon < x$.

35. Solve for all real values of x satisfying the inequality $x + 6 < |x|$. (*Hint:* Break up the solution into the cases where x is non-negative and x is negative. In the former case, we know that $|x| = x$; in the latter case $|x| = -x$. Try each case separately and see what they lead to.)

36. Solve for all real values of x such that $x^2 + 3 < 4|x|$.

37. Find all real values of x satisfying the relation

$$|x| < |x|^3$$

38. Find all real values of x satisfying the inequality $|x + 1| < 2|x|$.

§ 2.4

In the following problems, the concepts of finiteness, summation, and products of unspecified numbers of real numbers, exponents, and factorials occur. Although we have not given formal definitions of these things, their elementary properties may be assumed for the purpose of solving these problems.

39. Show by mathematical induction that the sum of the first n positive integers is $n(n + 1)/2$.
40. Prove inductively that $n < 2^n$ for each positive integer n.
41. Show by induction that if $x \in R$ and $1 < x$, then $1 < x^n$ for each positive integer n.
42. Show that if $x \in R$ and $1 < x$, that is, $x = 1 + h$, where h is positive, then $1 + nh < x^n$ for each positive integer n.
43. Prove inductively that $n^2 + n$ is divisible by 2 whenever $n \in P$.
44. Prove inductively that $n^3 - n$ is divisible by 6 whenever $n \in P$. (*Hint:* You may wish to use the result of Problem 43 at one step of the proof.)
45. Prove inductively that for each odd positive integer n, $n^2 - 1$ is divisible by 8. (*Hint:* Since to each odd positive integer n there corresponds a positive integer k such that $n = 2k - 1$, we consider $(2k - 1)^2 - 1 = 4k^2 - 4k$ and show inductively that this number is divisible by 8 whenever k is a positive integer.)
46. Prove that $2^n < n!$ for each positive integer $n \geq 4$, using mathematical induction. (*Hint:* Replace the given problem by that of showing $2^{k+3} < (k + 3)!$ for each positive integer k. This is a device that can be used to establish some statements that do not hold for all positive integers, but are valid for all positive integers that are greater than or equal to some fixed number.)
47. Prove that if $Q \subset P$, $c \in Q$, and $x + 1 \in Q$ whenever $x \in Q$, then Q contains all positive integers that are greater than or equal to c. (*Hint:* Define a new set $Q' = \{n \mid n \in P$ and there exists $x \in Q$ such that $n = x - c + 1\}$. Check that $1 \in Q'$ and that $n + 1 \in Q'$ whenever $n \in Q'$. Infer that $Q' = P$. Refer back to Q and draw the desired conclusion.)
48. Use the result of Problem 47 above to prove that $2n + 1 < 2^n$ holds for each positive integer $n \geq 3$.
49. Prove by induction that $n^2 < 2^n$ for each positive integer $n \geq 5$.
50. If $x \in T$, $y \in T$, and $0 < x < y$, then $0 < x^n < y^n$ for each positive integer n.
51. Prove inductively that if x and y are real numbers, then $x^n - y^n$ is divisible by $x - y$. (*Hint:* The polynomial P in two variables is divisible by a polynomial S, also in two variables, if and only if there exists a polynomial Q such that $P(x, y) = Q(x, y) \cdot S(x, y)$ for each $x \in R$ and each $y \in R$. In this problem, it is necessary to use a well-known device; namely, we write $x^{k+1} - y^{k+1} = x^k(x - y) + y(x^k - y^k)$. You may wish to use this. You may assume that sums and products of polynomials are again polynomials.)
52. Prove by induction that if $x \in R$ and $y \in R$, then $x^n + y^n$ is divisible by $x + y$ whenever n is an odd positive integer. (*Hint:* Use hint to Problem 51 above.)

53. Prove by induction that if $x \in R$ and $y \in R$, then

$$x^n - y^n = (x - y)(x^{n-1} + x^{n-2}y + x^{n-3}y^2 + \ldots + y^{n-1})$$

for each positive integer n. Thus show that if x and y are both non-negative then $x^n = y^n$ if and only if $x = y$.

54. Prove by induction that if n is any positive integer, then the product of n positive integers, each of which is greater than 1, exceeds n.

55. A positive integer is called a *prime number* if and only if it cannot be represented as a product of two or more positive integers each of which is greater than 1; otherwise it is called a *composite number*. The only exception is 1 itself, which is not considered to be a prime nor composite. Prove that every positive integer different from 1 may be represented as a product of finitely many prime numbers (possibly with repetitions). (*Hint:* This problem calls for the use of (ix).)

§§ 2.5, 2.6

56. Prove (i) of § 2.5.
57. Prove (ii) of § 2.5.
58. Prove (iii) of § 2.5.
59. Show that if x and x' are positive real numbers, then $xx'/(x + x')$ is positive and less than both of the numbers x and x'; if x and x' are rational, it is also rational. Similarly, if x, x', and x'' are positive, then $xx'x''/(xx' + x'x'' + x''x)$ is positive, and less than all three of the numbers x, x', and x''.
60. Use the technique of § 2.6 to show that $\sqrt{3}$ is not rational, but is a real number.
61. Prove that the sum, difference, and product of a rational with an irrational number are irrational.
62. Let c be a positive real number, n a positive integer,

$$S = \{y \mid y \in R, 0 \leq y, \text{ and } y^n < c\}$$
$$T = \{y \mid y \in R, 0 \leq y, \text{ and } c < y^n\}.$$

Show that: (a) S is non-empty and has no largest member, (b) T has no smallest member, (c) every member of T is an upper bound for S, (d) S has a least upper bound x that must satisfy the equation $x^n = c$. This number x is the one conventionally defined as $\sqrt[n]{x}$ or $x^{1/n}$. The equation $x^n = 0$ has only the solution $x = 0$, and so we define $0^{1/n} = 0$.

63. Show that if $0 \leq x < y$ and $n \in P$, then $x^{1/n} < y^{1/n}$.
64. If $0 \leq x$, $m \in P$ and $n \in P$, we define $x^{m/n} = (x^m)^{1/n}$. Show that $x^{m/n} = (x^{1/n})^m$. (*Hint:* Let $y = (x^m)^{1/n}$. Then $y^n = x^m$. Let $z = x^{1/n}$. Then $z^m = (x^{1/n})^m$; also $z^n = x$; $z^{mn} = x^m$. Use Problem 53 above to show $z^m = y$.)

§§ 2.7 to 2.14

65. Prove that if r is any given rational number then there exist positive integers m and n without common factors except 1, such that $r = m/n$. Also if m' and n' are any positive integers without common factors such that $r = m'/n'$ then $m = m'$, $n = n'$. (*Hint:* Let S denote the set of all denominators for the quotients of the admissible kind that are equal to r. Choose n as the smallest member of S. Use the result of Problem 55 above.)
66. Show that if $S' \subset S \subset R$, M is an upper bound of S, and m is a lower bound

of S, then M is necessarily an upper bound of S' and m is necessarily a lower bound of S'.

67. Prove that between any two different real numbers, whether they are rational or not, there exists an irrational number. (*Hint:* Use $\sqrt{2}$, which is known to be irrational, as your "building block.")

68. Let $A \neq \varnothing$ and $B \neq \varnothing$ be such subsets of R that $A \cup B = R$, $A \cap B = \varnothing$, and for all x and all y, if $x \in A$ and $y \in B$ then $x < y$. Show that there exists a real number z (which may belong to A or to B) such that if $z \neq x \in A$ and $z \neq y \in B$ then $x < z < y$. A partition of R into sets A and B possessing the properties named is called a *Dedekind cut*, and when Dedekind cuts are made in the rational number system, each such cut can be used as the definition of a real number. This is one way of defining the real number system from the rationals.

69. Let $x \in R$, $A = \{r \mid r$ is rational and $r < x\}$, and $B = \{r \mid r$ is rational and $x \leq r\}$. Show that A and B comprise a Dedekind cut in the rationals. Show that the real number determined by this cut in accordance with Problem 68 is x itself.

70. Let $A = \{r \mid r$ is rational and $r < 0$, or $0 \leq r$ and $r^2 \leq 7\}$, $B = \{r \mid r$ is rational, $0 \leq r$ and $7 < r^2\}$. Show that A and B comprise a cut in the rationals that determine the number conventionally called $\sqrt{7}$. (It might be noted that although B includes in its definition any positive rational s such that $s^2 = \sqrt{7}$, there is in fact no such rational number.)

71. Let S be a sequence of real numbers, that is, $S(n) \in R$ for each $n \in P$ (cf. Definition 4.1), such that $S(n) \leq S(n + 1) \leq M$ for each $n \in P$, where M is a fixed real number. Let

$$A = \{r \mid r \text{ is rational and there exists } n \in P \text{ such that } r < S(n)\}$$

$$B = \{r \mid r \text{ is rational and } S(n) \leq r \text{ for each } n \in P\}.$$

Show that A and B comprise a Dedekind cut in the rational numbers. Let x be the real number this cut defines. Show that $x = $ l.u.b. $\mathscr{R}(S)$.

3

Finite and Infinite Sets

Intuitively we regard a set S as being *finite* if we can take its elements one by one and count them using the positive integers 1, 2, 3, etc., with the process stopping at some positive integer, when the members of S have been exhausted. However, this procedure does more than merely determine that S is finite. In fact, the last integer required in the counting process tells us what is conventionally known as the *number of elements in S*. For our purposes, this information is usually not particularly important. Generally, we are interested only in whether or not a set is finite. We find it convenient to adopt a criterion for finiteness that is not quite so rigid as the one described heuristically above, one in which the final integer used may or may not tell us the number of elements in S. We drop the requirement that each element must be counted once only and allow the possibility that some or all members of S may be counted more than once. It is still intuitively clear that if there exists some means of counting the members of S in which all members of S are counted at least once, and if the counting terminates at some positive integer, then S is finite. We wish to formalize this idea, and it is with this in mind that we proceed.

3.1 Definition. If $n \in P$, then we agree that

$$\hat{n} = \{i \mid i \in P \text{ and } 1 \leq i \leq n\}.$$

3.2 Definition. A set S is *finite* if and only if $S = \varnothing$ or there exists a function f and a positive integer n such that $\mathscr{D}(f) = \hat{n}$ and $\mathscr{R}(f) = S$.

We see intuitively that as we run through the numbers 1, 2, ..., n, the values of f, namely $f(1), f(2), \ldots, f(n)$, run through all the members of S, with possible repetitions. Thus it is intuitively evident that f is our formal interpretation of the process of counting. Incidentally, if S is a non-empty finite set, it is also apparent that there must exist many functions f and positive integers n satisfying the requirements of Definition 3.2.

3.3 Lemma. If S is a finite set, then the set $Q = \{n \mid n \in P \text{ and there exists a function } f \text{ with } \mathscr{D}(f) = \hat{n}, \mathscr{R}(f) = S\}$ is non-empty if and only if S is non-empty.

Proof. If $S \neq \varnothing$, then $Q \neq \varnothing$ due to Definition 3.2. If $Q \neq \varnothing$, there exists $n \in P$ and a function f with $\mathscr{D}(f) = \hat{n}$, $\mathscr{R}(f) = S$. Then $1 \in \hat{n} = \mathscr{D}(f), f(1) \in \mathscr{R}(f) = S$, and $S \neq \varnothing$.

This lemma together with (vii) of § 2.4, justifies the following definition.

3.4 Definition. If S is any finite non-empty set, we define $\nu(S)$ as the smallest member of

$$\{n \mid n \in P \text{ and there exists a function } f \text{ with } \mathscr{D}(f) = \hat{n}, \mathscr{R}(f) = S\}.$$

If $S = \varnothing$, we define $\nu(S) = 0$.

3.5 Theorem. If S is a finite set, then $\nu(S)$ is a positive integer if and only if $S \neq \varnothing$; $\nu(S) = 0$ if and only if $S = \varnothing$.
Proof. This is an immediate consequence of Lemma 3.3, (vii) of § 2.4, and Definition 3.4.

3.6 Definition. For any finite set S, we define $\nu(S)$ as the *number of elements in S*, or the *cardinal number of S*.

This definition is not obviously the same as the one given heuristically in the opening remarks of this chapter, but it is equivalent to it, as will be shown by Theorem 3.30.

3.7 Definition. A set that is not finite is said to be *infinite*.

3.8 Theorem. If A and B are finite sets, then $A \cup B$ is a finite set.
Proof. If $A = \varnothing$, or $B = \varnothing$, or both are empty, then the statement of the theorem is clearly true, since $A \cup B$ reduces to A or B. In case $A \neq \varnothing$ and $B \neq \varnothing$, then by Definition 3.2, there exist functions f and g and positive integers m and n such that $\mathscr{D}(f) = \hat{m}$, $\mathscr{R}(f) = A$, $\mathscr{D}(g) = \hat{n}$, and $\mathscr{R}(g) = B$. We may let $p = m + n \in P$ and define the function h with $\mathscr{D}(h) = \hat{p}$ and

$h(i) = f(i) \qquad$ whenever $1 \leq i \leq m$ and $i \in P$

$h(i) = g(i - m) \qquad$ whenever $m + 1 \leq i \leq m + n = p$ and $i \in P$.

Clearly $\mathscr{D}(h) = \hat{p}$. We want to show that $\mathscr{R}(h) = A \cup B$. If $y \in \mathscr{R}(h)$ there exists $i \in \hat{p}$ such that $y = h(i)$. In case $1 \leq i \leq m$ then $y = h(i) = f(i) \in \mathscr{R}(f) = A \subset A \cup B$. In case $m + 1 \leq i \leq m + n = p$, which exhausts the possibilities, then $1 \leq i - m \leq n$, and so $y = h(i) = g(i - m) \in \mathscr{R}(g) = B \subset A \cup B$. Therefore $\mathscr{R}(h) \subset A \cup B$. On the other hand, if $y \in A \cup B$, then $y \in A$ or $y \in B$, so there exists a positive integer $i \leq m$ with $f(i) = y$, or there exists a positive integer $j \leq n$ with $g(j) = y$. In the former case, $i \in \hat{m}$ and $y = f(i) = h(i) \in \mathscr{R}(h)$. In the latter case, $m + 1 \leq m + j \leq m + n = p$, so that $y = g(j) = h(m + j) \in \mathscr{R}(h)$. Hence in either case $y \in \mathscr{R}(h)$, so $A \cup B \subset \mathscr{R}(h)$. Thus $\mathscr{R}(h) = A \cup B$ and $A \cup B$ is finite by Definition 3.2.

The theorem just proved can easily be extended directly to the unions of three sets, or four sets, etc. However, we shall prove a general theorem that covers these as special cases.

3.9   Theorem.   If $\mathscr{K}$ is a finite family of finite sets, then $\bigcup \mathscr{K}$ is a finite set.

Proof.   We shall prove this by induction on the number of elements in $\mathscr{K}$. For this purpose we let

$$Q = \{n \mid n \in P \text{ and } \bigcup \mathscr{F} \text{ is finite whenever}$$
$$\mathscr{F} \text{ is a family of finite sets and } \nu(\mathscr{F}) \leq n\}.$$

We consider first an arbitrary family $\mathscr{F}$ of finite sets with $\nu(\mathscr{F}) \leq 1$. By Theorem 3.5, either $\nu(\mathscr{F}) = 0$ or $\nu(\mathscr{F}) = 1$. By the same theorem, if $\nu(\mathscr{F}) = 0$ then $\mathscr{F} = \varnothing$, and $\bigcup \mathscr{F} = \varnothing$ is a finite set. In case $\nu(\mathscr{F}) = 1$, then by definition of $\nu(\mathscr{F})$, there exists a function f with $\mathscr{D}(f) = \hat{1}$, $\mathscr{R}(f) = \mathscr{F}$. It follows that $\mathscr{R}(f) = \langle f(1) \rangle$, and so by Theorem 1.22, $\bigcup \mathscr{F} = f(1) \in \langle f(1) \rangle = \mathscr{F}$. Consequently $\bigcup \mathscr{F}$ is a finite set since it is a member of $\mathscr{F}$. Thus we conclude that $1 \in Q$.

Next we assume that $n \in Q$ and prove therefrom that $n + 1 \in Q$. To this end, we take an arbitrary family $\mathscr{F}$ with $\nu(\mathscr{F}) \leq n + 1$. Of course, if $\nu(\mathscr{F})$ does not exceed n, then $\bigcup \mathscr{F}$ must be finite as a consequence of our assumption that $n \in Q$; hence we need consider only the case $\nu(\mathscr{F}) = n + 1$. We let $m = n + 1$. By definition of $\nu(\mathscr{F})$, there exists a function f with $\mathscr{D}(f) = \hat{m}$, $\mathscr{R}(f) = \mathscr{F}$. We let $g = (f \mid \hat{n})$, $\mathscr{R}(g) = \mathscr{G}$, and $\mathscr{H} = \langle f(n + 1) \rangle$. Since $g \subset f$ then $\mathscr{G} = \mathscr{R}(g) \subset \mathscr{R}(f) = \mathscr{F}$; also $\mathscr{H} \subset \mathscr{R}(f) = \mathscr{F}$; therefore $\mathscr{G} \cup \mathscr{H} \subset \mathscr{F}$. On the other hand, if $x \in \mathscr{F}$ there exists $i \in \hat{m}$ so that $x = f(i)$. In case $1 \leq i \leq n$ then $x = f(i) = g(i) \in \mathscr{G}$; if $i = n + 1$ then $x = f(n + 1) \in \mathscr{H}$; in any case $x \in \mathscr{G} \cup \mathscr{H}$ whence $\mathscr{F} \subset \mathscr{G} \cup \mathscr{H}$. Thus $\mathscr{F} = \mathscr{G} \cup \mathscr{H}$ and by Theorem 1.24,

$$(1) \qquad\qquad \bigcup \mathscr{F} = \left(\bigcup \mathscr{G}\right) \cup \left(\bigcup \mathscr{H}\right).$$

We consider

$$A = \{k \mid k \in P \text{ and there exists a function } h \text{ with } \mathscr{D}(h) = \hat{k}, \ \mathscr{R}(h) = \mathscr{G}\};$$

note that $n \in A$, and since $\nu(\mathscr{G})$ is by definition the smallest member of A, we conclude that $\nu(\mathscr{G}) \leq n$. Since $\mathscr{G}$ is a finite family of finite sets, we can now use our inductive assumption that $n \in Q$ to infer that $\bigcup \mathscr{G}$ is a finite set. By Theorem 1.22, $\bigcup \mathscr{H} = f(n + 1) \in \mathscr{F}$, hence $\bigcup \mathscr{H}$ is a finite set since all members of $\mathscr{F}$ are finite sets. From (1) and Theorem 3.8 we deduce that $\bigcup \mathscr{F}$ is a finite set.

This essentially completes the proof of our theorem. For if $\mathscr{K}$ is any finite family of finite sets, then by Theorem 3.5, either $v(\mathscr{K}) = 0$ and $\mathscr{K} = \varnothing$, or $v(\mathscr{K}) \in P$; and by what we have just proved, $\bigcup \mathscr{K}$ is finite in each of these cases.

3.10 Theorem. If A is a finite set and $B \subset A$, then B is a finite set.
Proof. If $B = \varnothing$, the theorem is surely true. Accordingly we consider the case $\varnothing \neq B \subset A$. Then there exists z such that $z \in B$.

Since $B \neq \varnothing$, it is clear that $A \neq \varnothing$, so there exists a function f and a positive integer n such that $\mathscr{D}(f) = \hat{n}$, $\mathscr{R}(f) = A$. We let T denote the subset of $\hat{n}$ for which $k \in T$ if and only if $f(k) \in B$. It follows that for each $i \in (\hat{n} - T)$,[1] $f(i) \in (\mathscr{R}(f) - B) = A - B$. We now define the function h with $\mathscr{D}(h) = \hat{n}$ so that

$$h(i) = f(i) \text{ whenever } i \in T,$$
$$h(i) = z \text{ whenever } i \in (\hat{n} - T).$$

We see that for each $i \in \hat{n}$, $h(i) \in B$, thus $\mathscr{R}(h) \subset B$. Also, if x is any element of B, there exists a positive integer $i \in T$ for which $x = f(i) = h(i) \in \mathscr{R}(h)$. Thus $B \subset \mathscr{R}(h)$, and we now see that $\mathscr{R}(h) = B$. Consequently h and n satisfy the requirements of Definition 3.2 and B is accordingly a finite set.

3.11 Corollary. If A is a finite set and B is any set, then $A \cap B$ and $A - B$ are finite sets.
Proof. $A \cap B \subset A$ and $(A - B) \subset A$, thus Theorem 3.10 applies.

3.12 Corollary. If $B \subset A$ and B is an infinite set, then A is an infinite set.
Proof. If A were a finite set, then by Theorem 3.10 B would necessarily be finite, contrary to assumption.

3.13 Corollary. If A is an infinite set, $B \subset A$, and B is a finite set, then $A - B$ is an infinite set.
Proof. $A = B \cup (A - B)$; and if $A - B$ were finite, then A would be finite by Theorem 3.8. contrary to assumption.

3.14 Corollary. If A is an infinite set and B is a finite set, then $A - B$ is an infinite set.
Proof. Since $A \cap B$ is a finite set by Corollary 3.11, and since $A - B = A - (A \cap B)$, where $A \cap B \subset A$, then by Corollary 3.13, it follows that $A - B$ is an infinite set.

3.15 Theorem. If f is a function, and $\mathscr{D}(f)$ is a finite set, then $\mathscr{R}(f)$ is a finite set.

---

[1] T is certainly non-empty, but it may happen that $\hat{n} - T = \varnothing$. The statement is still valid in this case, however, as an implication in which the antecedent is false.

Proof. If $\mathscr{D}(f) = \varnothing$, then there exist no ordered pairs $(x, y) \in f$, and so $f = \varnothing$. In this case $\mathscr{R}(f) = \varnothing$, so $\mathscr{R}(f)$ is finite.

If $\mathscr{D}(f) \neq \varnothing$, there exists a function g and a positive integer n such that $\mathscr{D}(g) = \hat{n}$, $\mathscr{R}(g) = \mathscr{D}(f)$. By Theorem 1.48, $f \circ g$ is a function with $\mathscr{D}(f \circ g) = \hat{n}$, $\mathscr{R}(f \circ g) = \mathscr{R}(f)$. Evidently $\mathscr{R}(f)$ is a finite set.

We now apply some of the ideas we have developed to sets of real numbers. This was the main purpose of the work done thus far.

3.16 Theorem. Each non-empty finite set of real numbers is bounded (above and below).

Proof. We will prove this by induction on the number of elements in the finite sets of real numbers. To this end, we let

$$Q = \{n \mid n \in P \text{ and if } A \subset R, 1 \leq v(A) \leq n, \text{ then } A \text{ is bounded}\}.$$

We take an arbitrary subset A of the real numbers with $v(A) = 1$. By definition of $v(A)$, there exists a function f with $\mathscr{D}(f) = \hat{1}$, $\mathscr{R}(f) = A$. It follows that $A = \mathscr{R}(f) = \langle f(1) \rangle$. We see that for any x, if $x \in A$, then $-|f(1)| \leq x \leq |f(1)|$. Hence A is bounded above by $|f(1)|$, below by $-|f(1)|$, and so we conclude that $1 \in Q$.

Next, we assume that $n \in Q$ and show therefrom that $n + 1 \in Q$. Accordingly, we take an arbitrary set $A \subset R$ with $1 \leq v(A) \leq n + 1$. If $v(A)$ does not exceed n, then by the assumption that $n \in Q$, we infer that A must be bounded above and below. Thus we need consider only case $v(A) = n + 1$. We let $m = n + 1$, and by virtue of the definition of $v(A)$, there exists a function f with $\mathscr{D}(f) = \hat{m}$, $\mathscr{R}(f) = A$. We let $g = (f \mid \hat{n})$, $B = \mathscr{R}(g)$, and observe that $B = \mathscr{R}(g) \subset \mathscr{R}(f) = A \subset R$. We also let $C = \langle f(n + 1) \rangle$, note that $C \subset \mathscr{R}(f) = A$ and conclude that $B \cup C \subset A$. Now if x is any element of A, then $x \in \mathscr{R}(f)$ and there exists $i \in \hat{m}$ such that $x = f(i)$. In case $1 \leq i \leq n$, then $f(i) = g(i)$ and so $x \in \mathscr{R}(g) = B \subset B \cup C$. In case $i = n + 1$, then $x = f(n + 1) \in \langle f(n + 1) \rangle = C \subset B \cup C$. Thus we infer that $A \subset B \cup C$, and so finally that $A = B \cup C$.

Next we consider $S = \{k \mid k \in P \text{ and there exists a function } h \text{ with } \mathscr{D}(h) = \hat{k}, \mathscr{R}(h) = B\}$. Since $\mathscr{D}(g) = \hat{n}$, $\varnothing \neq \mathscr{R}(g) = B \subset A \subset R$, then $n \in S$. Also since $v(B)$ is the smallest member of S by definition, then $1 \leq v(B) \leq n$. From our inductive assumption that $n \in Q$, we now infer that B is bounded. Thus there exists a positive number M' such that for all x, if $x \in B$ then $-M' \leq x \leq M'$.[2] Now let $N = M' + |f(n + 1)|$, and let

---

[2] If a set $A \subset R$ is bounded above and below, there exist numbers K and L such that, for each $x \in A$, $K \leq x \leq L$. If we let $M = |K| + |L|$, then clearly $-M \leq K \leq x \leq L \leq M$ holds for each $x \in A$. Thus when a set is bounded above and below, we can always replace the bounds by new bounds that are equal but opposite in sign. This is a convenience we will use frequently.

us consider an arbitrary element $x \in A$. Then either $x \in B$ or $x \in C$. If $x \in B$, we have $-M' \leq x \leq M'$, which ensures that $-N \leq -M' \leq x \leq M' \leq N$. If $x \in C$, then $-|f(n+1)| \leq x = f(n+1) \leq |f(n+1)|$, which also ensures that $-N \leq -|f(n+1)| \leq x \leq |f(n+1)| \leq N$. Thus it is clear that $-N$ and N serve as lower and upper bounds, respectively, for the set A. Consequently, $n+1 \in Q$ and so $Q = P$.

Now we consider any non-empty finite set $A \subset R$. By Theorem 3.5, $\nu(A) \in P$, whence by what we just proved, A is necessarily bounded and the proof is complete.

3.17　Corollary. Any non-empty set of real numbers that is unbounded either above or below is an infinite set.

3.18　Corollary. P is an infinite set.
Proof. By Corollary 2.8, P has no upper bound, hence Corollary 3.17 may be applied.

3.19　Corollary. If h is a positive real-valued function, $n \in P$, and $\mathscr{D}(h) = \hat{n}$, then there exists a positive number d such that $d < h(i)$ holds for each $i \in \hat{n}$.
Proof. We so define the function f that $\mathscr{D}(f) = \hat{n}$, $f(i) = 1/h(i)$ for each $i \in \hat{n}$, and we let $A = \mathscr{R}(f)$. Evidently A is a finite set of positive real numbers, hence by Theorem 3.16 A has an upper bound M' such that $0 < x \leq M'$ whenever $x \in A$. Putting $M = M' + 1$, we obtain the strict inequality $0 < x < M$ whenever $x \in A$. Thus for each $i \in \hat{n}$, we have $0 < f(i) = 1/h(i) < M$, and so $1/M < h(i)$ holds for each such i. Taking $d = 1/M$ completes the proof.

3.20　Theorem. If A is any finite non-empty set of real numbers, then A has a largest and a smallest member.
Proof. By Theorem 3.16, A is bounded above (below), thus A has a l.u.b. M (g.l.b. m). If $y = M$ ($y = m$) for some $y \in A$, then M is the largest (m is the smallest) member of A. We want to show that this must happen for some $y \in A$. For if not, then we must have $x < M$ ($m < x$) for each $x \in A$. Since A is finite and non-empty, there exists a positive integer n and a function f such that $\mathscr{D}(f) = \hat{n}$, $\mathscr{R}(f) = A$. For each $i \in \hat{n}$, we define $h(i) = M - f(i) > 0$ $(h(i) = f(i) - m > 0)$. Applying Corollary 3.19, we find d so that $0 < d < h(i)$ for each $i \in \hat{n}$. For an arbitrary element $x \in A$, there exists $i \in \hat{n}$ with $x = f(i)$, and evidently $0 < d < M - x = M - f(i) = h(i)$ $(0 < d < x - m = f(i) - m = h(i))$, whence $x < M - d$ $(m + d < x)$ holds for each $x \in A$. This contradicts the assumption that M is the l.u.b. (m is the g.l.b.) of A. Now $x \leq M$ ($m \leq x$) whenever $x \in A$, and since we have shown that the strict inequality cannot hold for each $x \in A$, there necessarily exists $y \in A$ such that $y = M$ ($y = m$) as desired.

3.21 Definition. If y is any real number and $I \subset R$, we shall say that I is a *neighborhood* of y if and only if there exists a positive number ϵ such that $I = \{x \mid y - \epsilon < x < y + \epsilon\} = \{x \mid |x - y| < \epsilon\}$.

If we represent the real numbers as points on a line in the manner of analytic geometry, then the neighborhood of the real number y just defined may be interpreted geometrically as the set of all points on the line that are within a distance ϵ of the point representing y.

We come now to one of the fundamental definitions of mathematical analysis.

3.22 Definition. If S is a set of real numbers and y is a real number, then y is said to be a *cluster point* or *limit point* or *point of accumulation* of S if and only if for each positive number ϵ, the set $S \cap \{x \mid y - \epsilon < x < y + \epsilon\} = S \cap \{x \mid |x - y| < \epsilon\}$ is infinite. We shall sometimes express this situation by saying that *every neighborhood of y contains infinitely many members of S.*

The following theorem gives a criterion that is sometimes easier to use than the definition itself in determining whether or not a particular real number is or is not a cluster point of a given set. After it has been established, we will consider some specific examples.

3.23 Theorem. If $S \subset R$ and $y \in R$, then y is a cluster point of S if and only if each neighborhood of y contains a member z of S different from y, that is, if and only if for each $\epsilon > 0$ there exists $z \in S$ such that $z \neq y$ and $|z - y| < \epsilon$.

Proof. If y is a cluster point of S, then for any $\epsilon > 0$, $S \cap \{x \mid |x - y| < \epsilon\}$ is infinite; also $\langle y \rangle$ is a finite set, so by Corollary 3.14,

$$(S \cap \{x \mid |x - y| < \epsilon\} - \langle y \rangle)$$

is an infinite set and therefore non-empty. Thus there exists an element z belonging to this last set and it clearly satisfies the stated requirements.

On the other hand, if y is not a cluster point of S, then there exists $\epsilon > 0$ such that $B = S \cap \{x \mid |x - y| < \epsilon\}$ is a finite set, and so also is $B - \langle y \rangle$. If $B - \langle y \rangle = \varnothing$, then there exists no element $z \in (B - \langle y \rangle)$, so that for this particular ϵ, there exists no z satisfying the requirements of the theorem. If $B - \langle y \rangle \neq \varnothing$, then we so define the function f on $B - \langle y \rangle$ that $f(x) = |x - y|$ whenever $x \in (B - \langle y \rangle)$. Since $B - \langle y \rangle = \mathscr{D}(f)$ is a finite set, then so is $\mathscr{R}(f) = \{u \mid \text{there exists } x \in (B - \langle y \rangle) \text{ such that } u = |x - y|\}$, by Theorem 3.15.

For any element $x \in (B - \langle y \rangle)$ we have $x \neq y$ and so $0 < |x - y|$, whence we see that $\mathscr{R}(f)$ is a finite set of positive real numbers, and it has a smallest member $v > 0$ by Theorem 3.20. We choose a positive number

η less than both ϵ and v (for instance, $\eta = \epsilon v/(\epsilon + v)$), and define $C = S \cap \{x \mid |x - y| < \eta\}$. Since $0 < \eta < \epsilon$, then $\{x \mid |x - y| < \eta\} \subset \{x \mid |x - y| < \epsilon\}$, whence $C \subset B$ and $(C - \langle y \rangle) \subset (B - \langle y \rangle)$. We consider an arbitrary element $x \in (C - \langle y \rangle)$. Then $x \in S$ and $|x - y| < \eta$. However, $x \in (B - \langle y \rangle)$ by virtue of the inclusion above, and since v is the smallest member of $\mathscr{R}(f)$, it follows that $0 < \eta < v \leq |x - y|$. Thus we infer that $|x - y| < |x - y|$, a contradiction leading to the conclusion that $C - \langle y \rangle = \varnothing$.

We have shown that if y is not a cluster point of S, then there exists some neighborhood of y, namely $\{x \mid |x - y| < \eta\}$, containing no members of S different from y, and this completes the proof of the theorem.

Definition 3.22 does not require that $y \in S$ in order that y be a cluster point of S. We will now give a few examples to help gain an idea of the nature of cluster points.

We take first $S = \{x \mid 0 < x \leq 1\}$. We will show that 0 is a cluster point of S, although $0 \notin S$. Consider an arbitrary positive number ϵ, and let $N = \{x \mid -\epsilon < x < \epsilon\}$. Clearly N is a neighborhood of 0. Now the positive number $\epsilon/(1 + \epsilon)$ is less than 1, and less than ϵ, so belongs to both N and S, and is different from 0. Consequently, by Theorem 3.23, 0 is a cluster point of S. We next show that 1 is a cluster point of S. Again we take an arbitrary positive number ϵ, and consider $N' = \{x \mid 1 - \epsilon < x < 1 + \epsilon\}$; N' is a neighborhood of 1. We let u denote the larger of the two numbers 0 and $1 - \epsilon$ (which may be negative), and note that $0 \leq u < 1$. If we take any number z such that $u < z < 1$, then it follows that $z \in S$, $z \in N'$, and $z \neq 1$, so that Theorem 3.23 may again be applied. It is not difficult to show that any number y such that $0 < y < 1$ is also a cluster point of S. On the other hand, we will show that the number $1\frac{1}{4}$ is not a cluster point of S. Consider $N'' = \{x \mid 1\frac{1}{8} < x < 1\frac{3}{8}\}$. This is surely a neighborhood of $1\frac{1}{4}$ containing no points of S at all; hence $1\frac{1}{4}$ is not a cluster point of S. In general, to show that a number y is not a cluster point of a set, it is enough to exhibit a single neighborhood of y that contains no members of the set, except possibly y itself.

We take up another example. This time we let $S = \{x \mid$ there exists $n \in P$ such that $x = 1/n\}$. We will show that 0 is a cluster point of S. To this end we take an arbitrary positive number ϵ, and consider the neighborhood $N = \{x \mid -\epsilon < x < \epsilon\}$ of 0. By the Archimedean principle, there exists a positive integer n such that $1 < n\epsilon$, that is, $0 < 1/n < \epsilon$. Then $1/n \in S \cap N$ and $0 \neq 1/n$; hence by Theorem 3.23, 0 is a cluster point of S. It is possible to show that S has no other cluster points.

Finally, we consider $S = \{x \mid x$ is a rational number$\}$. The rational numbers comprise only a portion of the real number system, yet we will show that all real numbers are cluster points of S. We take an arbitrary real number y and an arbitrary positive number ϵ, and consider the neighborhood $N = \{x \mid y - \epsilon < x < y + \epsilon\}$ of y. By Theorem 2.12, there exists a rational number r such that $y < r < y + \epsilon$. Clearly $r \in S \cap N$ and $r \neq y$, hence we may apply Theorem 3.23 to infer that y is a cluster point of S.

We turn now to the question of determining what kinds of sets do or do not have cluster points.

3.24   Theorem.   If S is a finite set of real numbers, then S has no cluster points.

Proof.   Consider any real number y and any neighborhood N of y. Then $N \cap S$ is a subset of S and so is finite. Thus y is not a cluster point of S. The theorem follows from the arbitrary nature of y.

We see now that only infinite sets may have cluster points, but not all infinite sets have them, as we may ascertain from a consideration of P itself. However, we now show that certain kinds of infinite sets have at least one cluster point. This is known as the *Bolzano-Weierstrass Theorem*.

3.25   Theorem.   If S is a bounded (above and below) infinite set of real numbers, then S has a cluster point (possibly many).

Proof.   Due to the boundedness of S there exists a positive number M such that $-M < x < M$ for each $x \in S$. We let

$$T = \{x \mid S \cap \{y \mid y < x\} \text{ is an infinite set}\}.$$

We consider an arbitrary number t, $t < -M$. Then $S \cap \{y \mid y < t\} = \varnothing$ since the members of S are all greater than $-M$. Thus $t \notin T$ whenever $t < -M$. Hence we see that for any x, if $x \in T$ then $-M \leq x$, whence $-M$ is a lower bound for T. Also, we see that $S \cap \{y \mid y < M\} = S$, since all members of S are less than M. But S is by hypothesis an infinite set, consequently $M \in T$ and $T \neq \varnothing$. Thus by Theorem 2.14, T has a g.l.b. m, and evidently $-M \leq m$.

Now let us take an arbitrary positive number ϵ, and use Theorem 2.13 to find $x \in T$ such that $m \leq x < m + \epsilon$. We let

$$S' = S \cap \{y \mid m - \epsilon \leq y < x\}$$

$$S'' = S \cap \{y \mid y < m - \epsilon\}$$

$$S''' = S \cap \{y \mid y < x\}.$$

Evidently $S''' = S' \cup S''$ since each member y of S''' satisfies either $y < m - \epsilon$ or $m - \epsilon \leq y < x$; conversely, if y satisfies one of these inequalities, it is a member of S'''. Since $x \in T$, then S''' is an infinite set; since $m - \epsilon$ is less than m, the greatest lower bound of T, then $(m - \epsilon) \notin T$

and so S'' is a finite set. By Corollary 3.14, $S' = S''' - S''$ is an infinite set. We let

$$S^{(iv)} = S \cap \{y \mid m - \epsilon \leq y < m + \epsilon\}.$$

Now $x < m + \epsilon$, thus it follows that $S' \subset S^{(iv)}$, and so by Corollary 3.12, $S^{(iv)}$ is an infinite set. Now $\langle m - \epsilon \rangle$ is a finite set; and applying Corollary 3.14 again, we conclude that

$$S \cap \{y \mid m - \epsilon < y < m + \epsilon\} = S^{(iv)} - \langle m - \epsilon \rangle$$

is an infinite set. However, taking account of the arbitrary nature of ϵ, this is the condition that m be a cluster point of S, and the proof is complete.

The development we have given to the subject of finite and infinite sets is quite adequate for our purposes, and nothing additional is necessary. However, we shall devote the remainder of this chapter to a continuation of these studies in order to give a deeper insight into the subject, and incidentally to show in Theorem 3.30 the equivalence of our definition for the number of elements in a finite set with the conventional one. Because this material has no effect upon the continuity of the following chapters, it can be omitted if desired.

3.26 Definition. If A and B are sets, then we say that A *is in one-to-one correspondence with* B, or that A *is equivalent to* B, or that A *has the same cardinal number as* B, if and only if there exists a one-to-one mapping (cf. Definition 1.43) of A onto B.

The term "cardinal number" is used in set theory as a measure of the magnitude or number of elements in a set. It is possible to define this concept so that for any given set whatever, its cardinal number is determined. We did this for finite sets in Definition 3.6, and it can also be done for infinite sets; the reason for our not doing so is explained in the next paragraph. As we have introduced the expression, we have said only that if two sets can be mapped one onto the other in one-to-one fashion then they have the same cardinal number, without saying what that is. If we were to define cardinality directly, this last would be a consequence of that definition. So far as finite sets are concerned, if A and B are sets with the same cardinal number according to Definition 3.6, then it follows from Theorem 3.30 and Theorem 3.27 that A is equivalent to B. The converse is true by virtue of the same theorems.

In order to define cardinality as such, it is necessary to have a kind of standard set to use as a measuring stick. For example, in the case of finite sets, we used the natural numbers 0, 1, 2, 3, etc., as our measure for the magnitudes of given sets. It is easy for us to comprehend intuitively that different finite sets may have different magnitudes, and that the measure of such magnitudes may reasonably be represented by natural numbers. It is not quite so obvious how, or even whether, one might differentiate between magnitudes of infinite sets, because the intuition becomes vague here; and it is even less obvious what one might use as a measuring device for representing these magnitudes. This last is a difficult matter that requires a deep excursion into the theory of sets, and so is not taken up here.

We will concern ourselves with the concept of equivalence only as a one-to-one correspondence between sets, which does not require that we define cardinality itself. We would all agree that two finite sets are equivalent if and only if their elements may be physically paired off one by one, with none left over from either set after the pairing has been completed. In the case of infinite sets, it is physically impossible to complete such a pairing. Nevertheless we can and do simply generalize what is intuitively clear to us in the finite case, and agree that any two sets are equivalent if and only if there exists a pairing of their elements in a theoretical sense. This is what Definition 3.26 expresses in formal language. Other definitions for the equivalence of sets may be conceived, but this one has proved best to capture the spirit of the concept, and leads to many useful and reasonable results, even though some of its consequences may seem startling to the novice upon first encountering them. We shall give some examples of these peculiarities at this point.

We let $Q = \{n \mid n$ is an even positive integer$\}$. Quite clearly Q is a proper subset of P, yet there exists a one-to-one mapping of P onto Q, namely the function f so defined on P that $f(n) = 2n$ whenever $n \in P$. Thus P is equivalent to Q according to our definition.

Again, let $Q' = \{n \mid n$ is a positive integral power of 10$\}$. We see that Q' is a proper subset of P, much smaller even than Q above, since Q' consists of the number 10, 100, 1000, etc., which are farther and farther apart as one goes out into them according to their size. Yet we can define the function g on P so that $g(n) = 10^n$ whenever $n \in P$, and it is easily seen that g maps Q' onto P in one-to-one fashion, whence P is equivalent to Q'.

It is easy to construct much more greatly attenuated subsets of P than those just considered, and again to prove that P is equivalent to each of them. In fact, it is proved in Chapter 6 that every infinite subset of P is equivalent to P.

At this point, one may well begin to wonder about the appropriateness of our definition of equivalence. It seems a gross violation of the intuition that a portion of any set can be equivalent to the whole. This is due to the fact that in our experience with finite sets, nothing of the sort ever does or can occur, as will be shown in Corollary 3.33. It is only when dealing with infinite sets that we run into the peculiar situation we have just faced.

In fact, this peculiarity is characteristic of infinite sets, and the existence of a one-to-one mapping between a set and one of its proper subsets is taken by some mathematicians as the definition of an infinite set. In spite of the oddity we have just observed, our definition has some attractive consequences, as we now see.

3.27 Theorem. If A, B, and C are sets, then
 (i) A is equivalent to A.
 (ii) If A is equivalent to B, then B is equivalent to A.
 (iii) If A is equivalent to B, and B is equivalent to C, then A is equivalent to C.

Proof. (i) If $A \neq \varnothing$ we use the identity function I, so defined on A that $I(x) = x$ whenever $x \in A$. It is easily seen that I is a one-to-one mapping of A onto itself. If $A = \varnothing$, the null function yields vacuously the desired mapping.

 (ii) If A is equivalent to B, then there exists a function f mapping A onto B in one-to-one fashion. By Corollary 1.45, f^{-1} is a function mapping B onto A in one-to-one fashion, making B equivalent to A.

 (iii) If A is equivalent to B and B is equivalent to C, then there exist functions f and g mapping A onto B and B onto C, respectively, in one-to-one fashion. But then by Theorem 1.48, $g \circ f$ is a function mapping A onto C in one-to-one fashion, making A equivalent to C.

3.28 Lemma. If m and n are unequal positive integers, then $\hat{m}$ is not equivalent to $\hat{n}$.

Proof. We let T be the set of positive integers p for which there exists a positive integer k such that $p < k$ and $\hat{p}$ is equivalent to $\hat{k}$. Our lemma will be proved if we can show $T = \varnothing$. We propose to assume that $T \neq \varnothing$ and derive a contradiction.

Accordingly, T has a smallest member N and there exists a positive integer M such that $N < M$, and a one-to-one mapping f of N onto M. We observe that $N \neq 1$; for if $\mathscr{D}(f) = \hat{1}$ and $\mathscr{R}(f) = \hat{M}$ where $1 < M$, then $2 \leq M$, $1 \in M$, and $2 \in M$. This requires $(1, 1) \in f$ and $(1, 2) \in f$, contradicting the fact that f is a function. We let $N' = N - 1$, $M' = M - 1$ and note that $N' \in P$, $M' \in P$, and $N' < M'$. There exists $q \in \hat{N}$ such that $f(q) = M$, and we complete our proof by considering two possible situations.

CASE 1. $q = N$.
We let $g = (f \mid N')$. Clearly $\mathscr{D}(g) = \hat{N}'$ and it follows readily from the one-to-one nature of f that g is a one-to-one mapping with $\mathscr{R}(g) = \hat{M}'$. However, this makes N' a member of T contrary to the fact that N is the smallest member of T.

CASE 2. $1 \leq q \leq N'$.

Here we define h so that $\mathscr{D}(h) = \hat{N}'$, $h(i) = f(i)$ if $q \neq i \in N'$, $h(q) = f(N)$. It is easily checked from the one-to-one nature of f that $\mathscr{R}(h) = \hat{M}'$. We take i and j in $\hat{N}'$, $i \neq j$. If both i and j are different from q, then $h(i) = f(i) \neq f(j) = h(j)$ because f is a one-to-one mapping. If one of the numbers i or j equals q, for example if $j = q$, then since $i \neq N$ we have $h(i) = f(i) \neq f(N) = h(q) = h(j)$. Thus h is a one-to-one mapping of $\hat{N}'$ onto $\hat{M}'$, $N' \in T$, and we have the same contradiction we derived under Case 1.

3.29 Lemma. If A is a non-empty finite set, $n \in P$, and A has n members then every mapping of $\hat{n}$ onto A is one-to-one.

Proof. We consider an arbitrary function f such that

$$\mathscr{D}(f) = \hat{n}, \qquad \mathscr{R}(f) = A.$$

If $n = 1$, i and j belong to $\hat{1}$, and $f(i) = f(j)$, then we must have $i = j = 1$ since $\hat{1} = \langle 1 \rangle = \mathscr{D}(f)$. Thus f is a one-to-one mapping.

If $1 < n$, then $m = n - 1 \in P$. Suppose, if possible, that there exist positive integers i_0, j_0 in $\hat{n}$ such that $i_0 \neq j_0$ and $f(i_0) = f(j_0)$. We may assume $i_0 < j_0$ and define the function g on $\hat{m}$ so that for each $i \in \hat{m}$,

$$g(i) = f(i) \qquad \text{if } 1 \leq i < j_0,$$
$$g(i) = f(i + 1) \qquad \text{if } j_0 \leq i \leq m = n - 1.$$

It should be noted that in case $j_0 = n$, there is no positive integer i satisfying the conditions of the second line, in which case the definition of g is complete using only the first line.

Since $g(i) \in \mathscr{R}(f)$ for each $i \in \hat{m}$, then $\mathscr{R}(g) \subset \mathscr{R}(f)$. For each $x \in \mathscr{R}(f)$ there exists $i \in \hat{n}$ with $x = f(i)$. If $i = j_0$, then $x = f(j_0) = f(i_0)$ and $1 \leq i_0 < j_0 \leq n$, hence $x = f(i_0) = g(i_0) \in \mathscr{R}(g)$. If $i < j_0 \leq n$ then $x = f(i) = g(i) \in \mathscr{R}(g)$. Finally, if $j_0 < i \leq n$, then $j_0 \leq i - 1 \leq n - 1 = m$ and $x = f(i) = g(i - 1) \in \mathscr{R}(g)$. Thus $\mathscr{R}(f) \subset \mathscr{R}(g)$ and so $\mathscr{R}(f) = \mathscr{R}(g) = A$. However, $\mathscr{D}(g) = \hat{m}$ whence from Definition 3.4 $v(A) \leq m = n - 1$, contrary to the assumption that $v(A) = n$. Thus f must be a one-to-one mapping.

We now prove a theorem that establishes the logical equivalence of Definition 3.6 and the usual one for the cardinal number of a non-empty finite set.

3.30 Theorem. If A is a non-empty finite set and $n \in P$, then A has n elements if and only if A is equivalent to $\hat{n}$.

Proof. We let $v(A) = p$. According to Definition 3.4, there is a mapping of $\hat{p}$ onto A that must be one-to-one by Lemma 3.29, so $\hat{p}$ is equivalent to A. It follows that if A has n elements, $\hat{n}$ is equivalent to A. On the other hand, if $\hat{n}$ is equivalent to A, then $\hat{p}$ is equivalent to $\hat{n}$ by Theorem 3.27, whence $p = n$ by Lemma 3.28.

3.31 Corollary. If $n \in P$ then $\hat{n}$ has n members.
Proof. The identity mapping I defined by $I(i) = i$ for each $i \in \hat{n}$ is a one-to-one mapping of $\hat{n}$ onto itself and Theorem 3.30 applies.

3.32 Lemma. If A and B are finite sets with the same number of members, then every mapping of A onto B is one-to-one.
Proof. If $v(A) = v(B) = 0$ then $A = B = \varnothing$ due to Theorem 3.5. If f is any mapping of A onto B then $\mathscr{D}(f) = \mathscr{R}(f) = \varnothing$ and therefore $f = \varnothing$. It is easily checked that $\varnothing$ is vacuously a one-to-one mapping.

If $v(A) = v(B) = n \in P$, there exists a one-to-one mapping g of $\hat{n}$ onto A by Theorem 3.30. We let f be an arbitrary mapping of A onto B. Then $f \circ g$ is a mapping of $\hat{n}$ onto B, which is necessarily one-to-one by Lemma 3.29. Thus $(f \circ g) \circ g^{-1}$ is a one-to-one mapping of A onto B by Theorem 1.48. Since $((f \circ g) \circ g^{-1})(x) = f(g(g^{-1}(x))) = f(x)$ for each $x \in A$ due to Corollary 1.45, then $f = (f \circ g) \circ g^{-1}$ and the proof is complete.

3.33 Corollary. If A is a finite set and B is a proper subset of A, then B is not equivalent to A.
Proof. If $B = \varnothing$, then $A \neq \varnothing$ and if B were equivalent to A there would exist a function f with $\mathscr{D}(f) = A \neq \varnothing$, $\mathscr{R}(f) = \varnothing$, an impossibility.

If $B \neq \varnothing$ there exists $y \in B$. We define g on A so that

$$g(x) = x \qquad \text{if } x \in B$$
$$g(x) = y \qquad \text{if } x \in (A - B).$$

Since $A - B \neq \varnothing$, there exists $z \in (A - B)$, so that $z \neq y \in B$. Also $g(y) = y$ and $g(z) = y$ from our definition of g and so g is not a one-to-one mapping of A onto B. Hence by Lemma 3.32, A and B do not have the same number of elements. From Theorems 3.30 and 3.27 and Lemma 3.28 it follows that A cannot be equivalent to B.

3.34 Corollary. If A and B are finite sets, then $v(B) < v(A)$ if and only if B is equivalent to a proper subset of A and A is not equivalent to any subset of B.
Proof. We let $n = v(B)$, $m = v(A)$ and assume $n < m$. There exist functions f from $\hat{m}$ onto A and g from $\hat{n}$ onto B, both one-to-one. We let $h = (f \mid \hat{n})$, $A' = \mathscr{R}(h) \subset A$. The composite function $h \circ g^{-1}$ maps B onto A' in one-to-one fashion, and so from Theorems 3.27 and 3.30 we infer that $v(B) = v(A')$. If $A = A'$, then we must have $v(B) = v(A) = m$, a contradiction; therefore A' is a proper subset of A. On the other hand if A were equivalent to a subset B' of B, then by similar reasoning we must

have $v(A) = v(B') \leq v(B)$, again a contradiction which proves this part of the corollary.

Next we suppose that B is equivalent to a proper subset A' of A; then $v(B) = v(A') \leq v(A)$. If it were possible that $v(A') = v(A)$ then we must have A and A' both equivalent to $\hat{m}$, and so A must be equivalent to A', contrary to Corollary 3.33. Hence $v(B) < v(A)$ as we wanted to prove.

We will devote the rest of this chapter to establishing relationships between the cardinal members of the set P of positive integers, the set W of whole numbers, the set Ra of rationals, and the set R of real numbers.

3.35 Definition. Any set that is equivalent to P is said to be *denumerably* or *countably* infinite. A set that is finite or countably infinite is said to be *countable*.

3.36 Theorem. *W is countably infinite.*
Proof. We take the function F so defined on P that

$$F(n) = (n - 1)/2 \text{ if } n \text{ is an odd positive integer}$$

$$F(n) = -n/2 \text{ if } n \text{ is an even positive integer.}$$

Clearly $\mathscr{R}(F) \subset W$. We take an arbitrary element $k \in W$. If $k \in P$ then $n = 2k + 1 \in P$ and $F(n) = (n - 1)/2 = k$; $k \in \mathscr{R}(F)$; if $k = 0$, then $F(1) = 0 = k$, and again $k \in \mathscr{R}(F)$; finally if $-k \in P$ then $n = -2k \in P$ and $F(n) = -n/2 = k$, so that $k \in \mathscr{R}(F)$. Hence $W \subset \mathscr{R}(F)$ and so $\mathscr{R}(F) = W$.

It remains to be shown that F is a one-to-one mapping of P onto W. To this end we consider positive integers $m, n, m \neq n$. If both are even then $F(m) = -m/2 \neq -n/2 = F(n)$; if both are odd, $F(m) = (m - 1)/2 \neq (n - 1)/2 = F(n)$. If one is odd and the other is even, say m is odd and n is even, then there exist positive integers k and p such that $m = 2k - 1$, $n = 2p$; $F(m) = k - 1$ and $F(n) = -p$. Thus $F(n) < 0 \leq F(m)$ and so $F(m) \neq F(n)$. Thus F is a one-to-one mapping and the theorem is proved.

3.37 Lemma. $P \times P$ *is countably infinite.*
Proof. We so define the function G that for each ordered pair $(m, n) \in P \times P$,

$$G(m, n) = \frac{(m + n - 2)(m + n - 1)}{2} + m.$$

We note that the product of the two consecutive integers $m + n - 2$ and $m + n - 1$ is either 0 or an even positive integer; hence division by 2 leaves a value of zero or a positive integer, and hence we see that $G(m, n) \in P$, that is, $\mathscr{R}(G) \subset P$.

We consider an arbitrary element $j \in P$. We let S denote the set of positive integers p such that $j \leq (p - 1)p/2$. $S \neq \varnothing$ since $j + 1 \in S$, hence S has a smallest member k. Either $k - 1 = 0$, or $k - 1 \in P$, but in either case $k - 1 \notin S$ so that

$$0 \le \frac{(k-2)(k-1)}{2} < j \le \frac{(k-1)k}{2}$$

Also, it follows that $(k - 2)(k - 1)/2$ is a positive integer or zero. We define

$$(1) \qquad m = j - \frac{(k-1)(k-2)}{2},$$

whence m is a positive integer. We also define $n = k - m$. Upon substituting in (1) and reducing we obtain

$$(2) \qquad n = \left(\frac{k(k-1)}{2} - j\right) + 1 \ge 1$$

so that $n \in P$. Putting $k = m + n$ in (1) and solving for j we obtain $j = ((m + n - 2)(m + n - 1)/2) + m$, whence $j = G(m, n) \in \mathscr{R}(G)$. Hence $P \subset \mathscr{R}(G)$, and so $\mathscr{R}(G) = P$.

We complete our proof by showing that G is a one-to-one mapping of $P \times P$ onto P. To this end, we assume that $(m, n) \in P \times P$, $(m', n') \in P \times P$, and $G(m, n) = G(m', n')$. We let $q = m + n$, $q' = m' + n'$. Then

$$(3) \quad G(m, n) = \frac{(q-2)(q-1)}{2} + m, \quad G(m', n') = \frac{(q'-2)(q'-1)}{2} + m'.$$

We consider two possible cases.

CASE 1. $q \ne q'$. We may assume that $q < q'$, whence $q' \ge q + 1$. Subtracting in (3) we obtain

$$(4) \qquad m - m' = \frac{(q'-2)(q'-1) - (q-2)(q-1)}{2}$$

$$\ge \frac{(q-1)q - (q-2)(q-1)}{2} = q - 1,$$

and therefore $m \ge m' + q - 1$. However, $m' \ge 1$; therefore $m \ge q = m + n$, an impossibility since $n \in P$. Thus Case 1 does not arise.

CASE 2. $q = q'$. From (4) it follows that $m - m' = 0$, and since $m + n = q = q' = m' + n'$, we now see that $m = m'$, $n = n'$, and therefore $(m, n) = (m', n')$.

It is now clear that G is a one-to-one mapping and the proof is complete.

The inverse function G^{-1} gives a one-to-one mapping of P onto $P \times P$. It is the one usually given informally in pictorial fashion in textbooks, according to the scheme shown below, in which it is intuitively clear that each ordered pair $(m, n) \in P \times P$ appears exactly once, at the intersection of the mth row and the nth column. The ordered pairs are counted one by one by proceeding diagonally following the arrows. The count associated with the first few entries is shown parenthetically above the ordered pairs in question. It is intuitively clear that by this scheme each ordered pair is counted once and only once.

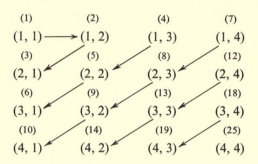

When counting out to the location of a given ordered pair (m, n), one may think of going out as far as the last full diagonal line that has to be counted to reach (m, n) and then going part of the way down the next diagonal to (m, n) itself. It is easily seen that the count along this diagonal will be exactly m. One may notice that the sums of the digits in the ordered pairs belonging to each diagonal line are constant. Counting the vertex $(1, 1)$ as the first diagonal line, it is seen that any given ordered pair (m, n) lies on the $(m + n - 1)$th diagonal. Thus there are $m + n - 2$ full diagonals to be counted before reaching (m, n). It is clear that the number of ordered pairs on all the diagonals up to and including the $(m + n - 2)$th is $1 + 2 + 3 + \ldots + (m + n - 2) = (m + n - 2)(m + n - 1)/2$. This is the first term in our expression for $G(m, n)$, and the added term "m" takes care of the final diagonal, as noted above. This is the intuitive motivation for our choice of G. Although the heuristic geometric thinking that led us to find G was nearly indispensable, it should be noted that it would be totally improper to consider such methods as "proofs" of anything. That is why, in establishing the properties of G in Lemma 3.37, we used strictly analytical methods, without reference to or dependence upon heuristic considerations.

3.38 Theorem. The set of all rational numbers is countably infinite.
Proof. We let $S = (P \times P) \cap \{(m, n) \mid m \text{ and } n \text{ have no common factors}$ but $1\}$. It is possible to show that for each positive rational number r, there exists one and only one ordered pair $(m, n) \in S$ such that $r = m/n$ (cf. Problem 65 in the Exercises at the end of Chapter 2). We shall assume

this to be known. This assumption is equivalent to the statement that there exists a one-to-one mapping J of the positive rationals onto S.

We now define $H = (G \mid S)$, where G is the function defined in Lemma 3.37. We let $P' = \mathscr{R}(H)$ and note that $P' \subset P$. Then $H \circ J$ maps the positive rationals onto P' in one-to-one fashion. Now we so define the function K on the set Ra of rationals that for an arbitrary rational r,

$$K(r) = (H \circ J)(r) > 0 \qquad \text{if } 0 < r$$
$$K(r) = -(H \circ J)(r) < 0 \qquad \text{if } r < 0$$
$$K(0) = 0.$$

We let $\mathscr{R}(K) = W'$. Since $\mathscr{R}(H \circ J) \subset P$, it is evident that $W' \subset W$. Since $H \circ J$ is a one-to-one mapping, it follows that K is a one-to-one mapping of Ra onto W'. Now $P \subset Ra$, therefore Ra is an infinite set by Corollary 3.12; consequently so is W'. By Theorem 3.36 there exists a one-to-one mapping of W onto P, and under this mapping, W' corresponds to a subset of P that must again be infinite. According to Corollary 6.4 (Chapter 6) this subset is countably infinite; hence by Theorem 3.27, W' and Ra must be countably infinite.

All the infinite sets considered up to this point have turned out to be equivalent to P, ranging from some highly attenuated subsets of P itself to the set of all rationals, which includes P and might well have been thought to be of a greater order of magnitude than P. One might well begin to question whether or not there exist any infinite sets that are "bigger" than P, that is, of greater cardinal number than P. Actually, we have given no formal definition for what such a statement might mean because it would take us too far afield. For our purposes, we will be content with something less than this. If Q, S, and T are sets and $Q \subset S \subset T$, it would seem reasonable that if there exists no one-to-one mapping of S onto Q, then T should have a greater cardinal number than Q. This is the point of view we adopt; it would be a consequence of a suitable general definition for comparison of cardinal numbers. On this basis we will show that the real numbers have a cardinal number exceeding that of P. This result is of great interest in itself, but the remarkable technique used in its proof is of even greater interest. It was devised by G. Cantor, one of the pioneers of set theory, and it has become known universally as *Cantor's diagonal process*. It has important applications in other connections in the theory of sets.

Before giving its proof, we have to make an assumption about real numbers that is a little too involved to prove at this point, although it is one with which everyone is familiar. We assume that every real number x from 0 to 1, not including 1, can be represented as an unending decimal,[3]

---

[3] This involves definition by induction and sums of infinite series.

that is, $x = .a_1 a_2 \ldots a_n \ldots$ where each of the digits $a_1, a_2, \ldots, a_n, \ldots$ is an integer between 0 and 9 inclusive. Some real numbers may be represented in two different ways, for example, $.5000 \ldots$ and $.4999 \ldots$ each represent $\frac{1}{2}$. If we disallow all decimals terminating in an unending string of nines, as we shall now agree to do, then the representations are unique, that is, there is a one-to-one correspondence between the set of all reals from 0 to 1, not including 1, and the set of all such decimals. For two such decimals to be different, it is sufficient that they differ in only one place. To show that R has a greater cardinal number than P, it will be sufficient to show that there exists no one-to-one correspondence between the set D of such decimals and P. Sets of this kind that cannot be counted with the positive integers are said to be *non-denumerably* or *uncountably infinite*.

3.39 Theorem. There does not exist a one-to-one correspondence between P and D.

Proof. To show the impossibility of this, we assume that there exists a one-to-one function f mapping P onto D. Then for each $i \in P$, $f(i)$ is an unending decimal that we may write as

$$f(i) = .a_{i1}a_{i2}a_{i3} \ldots a_{in} \ldots .$$

We now define b_i for each $i \in P$ so that

$$b_i = a_{ii} + 1 \quad \text{if } 0 \le a_{ii} \le 4, \qquad b_i = a_{ii} - 1 \quad \text{if } 5 \le a_{ii} \le 9$$

We now consider the decimal

(1) $$.b_1 b_2 \ldots b_n \ldots .$$

Clearly, for each $i \in P$, $b_i \ne a_{ii}$ and $b_i \ne 9$ due to our definition. Thus the decimal in (1) does not terminate in a string of nines, so belongs to D. Since we have assumed that $\mathscr{R}(f) = D$, then there exists some positive integer j such that

$$f(j) = .b_1 b_2 \ldots b_n \ldots .$$

Now also

(2) $$f(j) = .a_{j1}a_{j2} \ldots a_{jn} \ldots .$$

But if these values are to be the same, then the digits appearing in each place of (1) and (2) must be equal. In particular, though, $b_j \ne a_{jj}$ due to our construction, so that these decimals differ in at least one place and cannot be the same. This contradiction shows that our assumption that there exists a mapping function of the required type is false, and completes our proof.

The term "diagonal process" comes about naturally when one uses the following pictorial device for showing that the decimal (1) does not coincide with any of those represented by $f(i)$, $i \in P$. We imagine they are all listed:

$$f(1) = .a_{11}a_{12}a_{13} \ldots a_{1n} \ldots$$
$$f(2) = .a_{21}a_{22}a_{23} \ldots a_{2n} \ldots$$
$$f(3) = .a_{31}a_{32}a_{33} \ldots a_{3n} \ldots$$

$$\begin{matrix} \cdot & \cdot & \cdot & \cdot & & \cdot \\ \cdot & \cdot & \cdot & \cdot & & \cdot \\ \cdot & \cdot & \cdot & \cdot & & \cdot \end{matrix}$$

Now we imagine comparing each of these in turn with (1); since $b_1 \neq a_{11}$, then (1) is different from $f(1)$; since $b_2 \neq a_{22}$, then (1) is different from $f(2)$, etc. Thus (1) differs from all entries $f(1), f(2), f(3)$, etc; yet this list was supposed to contain all members of D, and in particular (1) itself. In effect, this is what we showed above. The diagonal numbers a_{11}, a_{22}, a_{33}, etc., are the ones that were used to show that (1) differs from each of the listed numbers.

Exercises

§§ 3.1 to 3.15

Either prove the truth of the first six problems, or show that they are not always true by giving explicit examples of the sets and families for which they are false.

1. If A and B are infinite sets, then $A \cap B$ is an infinite set.
2. If A is an infinite set and B is an infinite subset of A, then $A - B$ is necessarily a finite subset of A.
3. If A, B, C are sets and $A \cup B \cup C$ is infinite, then not more than one of them is a finite set.
4. If A is an infinite set and $x \in A$, then $A - \langle x \rangle$ is an infinite set.
5. If $\mathscr{F}$ is an infinite family of infinite sets, then $\bigcup \mathscr{F}$ is necessarily an infinite set.
6. If $\mathscr{F}$ is an infinite family of finite sets, then $\bigcup \mathscr{F}$ is necessarily an infinite set.
7. If A, B, C and D are finite sets, show that $A \cup B \cup C \cup D$ is finite using only Theorem 3.8. Prove the same thing by constructing a family $\mathscr{F}$ with the given sets as members and use Theorem 3.9.
8. If f is a function and $\mathscr{R}(f)$ is an infinite set, then $\mathscr{D}(f)$ is an infinite set. Use this to show that if A is an infinite set, and there exists a one-to-one mapping of A onto a set B, then B is an infinite set.
9. Prove that if A and B are finite sets, then $v(A \cup B) \leq v(A) + v(B)$. (*Hint:* Note proof of Theorem 3.8 and recall the definition of $v(A \cup B)$.)
10. If A, B, and C are such sets that $(A \cup B) - C$ is an infinite set, then either $A - C$ or $B - C$ is an infinite set; if $(A \cap B) - C$ is an infinite set, then both $A - C$ and $B - C$ are infinite sets.
11. Let A be such a set that for each $n \in P$ and each function f with $\mathscr{D}(f) = \hat{n}$, there exists x such that $x \in A$ and $x \notin \mathscr{R}(f)$. Show that the stated condition is both necessary and sufficient that A be an infinite set.

12. If f and g are functions, $\mathscr{D}(g) \subset \mathscr{R}(f)$, and $\mathscr{D}(f)$ is a finite set, then $\mathscr{R}(g)$ is a finite set. In particular, $\mathscr{R}(g \circ f)$ is a finite set.

13. If $A \neq \varnothing$ is a finite set, and u is a set or other mathematical object, show by actually defining a mapping function that $A \times \langle u \rangle$ is a finite set. (Recall Definition 1.34.)

14. Prove that if A and B are finite sets, then $A \times B$ is a finite set. (*Hint:* Use Problem 13 and induction based on the number of elements in B.)

15. Show that if A and B are sets and $A \times B$ is a finite set, then A and B are not necessarily both finite.

16. Show that the set of all rational numbers is infinite.

17. Show that if A is a finite set and there exists a one-to-one mapping of A onto a set B, then $v(A) = v(B)$.

18. Use Problem 8 to show that the following sets are infinite:

$$S = \{n \mid \text{there exists } m \in P \text{ for which } n = 1/m\}.$$
$$T = \{n \mid \text{there exists } m \in P \text{ for which } n = (m - 1)/m\}.$$
$$V = \{n \mid \text{there exists } m \in P \text{ for which } n = (-1)^m 2^m/(1 + 2^m)\}.$$

§§ 3.16 to 3.25

19. Given $S = \{x \mid 0 \leq x < 1\}$, show that if $0 < y < 1$ then y is a cluster point of S. Show that if $y < 0$ or $1 < y$ then y is not a cluster point of S.

20. If $S = \{x \mid \text{there exists } n \in P \text{ such that } x = 1/n\}$, show that 0 is the only cluster point of S.

21. Given $S = \{x \mid x \text{ is irrational and } 0 < x < 1\}$, prove that 0, 1, and every number between 0 and 1 are cluster points of S. (*Hint:* Use the result of Problem 67 in the Exercises at the end of Chapter 2.)

22. Let $S = \{x \mid \text{there exists } n \in P \text{ such that } x = (\frac{1}{2})^n\}$. Show that 0 is the only cluster point of S. (*Hint:* Use the result of Problem 40 in the Exercises at the end of Chapter 2.)

23. Prove that P has no cluster points.

24. Let S be a set of real numbers, $k \neq 0$ a real number, and let $T = \{x \mid \text{there exists } s \in S \text{ such that } x = ks\}$. Show that for each cluster point y of S, there is a cluster point ky of T, and conversely.

25. Given that u and v are real numbers, $u < v$, and $S = \{x \mid x \text{ is rational and } u < x < v\}$, show that u, v and all real numbers between u and v are cluster points of S.

26. Suppose that S is a set of real numbers with the following properties: (1) $s \in S$, $t \in S$, $s < t$; (2) if $x \in S$, $y \in S$ and $x < y$ then there exists $z \in S$ such that $x < z < y$. Show that S is necessarily an infinite set and must have a cluster point (possibly many).

27. Let $S = \{x \mid \text{there exists } n \in P \text{ such that } x = (-1)^n/n^2\}$. Prove that 0 is the only cluster point of S.

28. Let $S = \{x \mid \text{there exists } n \in P \text{ such that } x = (-1)^n(1 - 2^{-n})\}$. Show that 1 and -1 are the only cluster points of S.

§§ 3.26 to 3.27

29. Let $S = \{n \mid n \in P \text{ and } n \geq 10\}$. Show by finding a suitable mapping function that S is equivalent to P.

30. Let $S = \{x \mid \text{there exists } n \in P \text{ such that } x = (-1)^n/n\}$. Show that S is equivalent to P.

31. If $S = \{x \mid$ there exists $n \in P$ such that $x = n^n\}$, prove that S is equivalent to P.

32. Show that if A and B are sets each equivalent to P then $A \cup B$ is equivalent to P. (*Hint:* Find a function on P that runs through all values in A when odd positive integers are used, and runs through all values in B when even positive integers are used, similar to the technique of Theorem 3.36.)

33. Use the result of Problem 32 to show that if $\mathscr{F}$ is any non-empty finite family of sets, each equivalent to P, then $\bigcup \mathscr{F}$ is equivalent to P. Use induction on the number of members of $\mathscr{F}$.

§§ 3.28 to 3.39

34. Prove that the set of all irrational numbers is uncountably infinite. (*Hint:* Assume the contrary and arrive at a contradiction.)

35. Prove that the sets $P \times (P \times P)$ and $(P \times P) \times P$ are countably infinite.

36. If we define P^n so that $P^1 = P$ and $P^{n+1} = P^n \times P$ for each $n \in P$, show that P^n is countably infinite for each such n. The process of definition by induction used here may be assumed legitimate.

4

Sequences and Convergence
of Real-Valued Sequences

We shall devote this chapter to the study of *sequences*, with particular attention to numerical-valued sequences and to their limit theory.

4.1 Definition. An *infinite sequence* is a function whose domain is P. A *finite sequence* is a function whose domain is a set of the form $\hat{n}$, where $n \in P$ (cf. Definition 3.1).

Intuitively, the values of a sequence, finite or infinite, may be thought of as endowed with the same order relation as the positive integers, namely,

$$f(1), f(2), f(3), \ldots, f(n), \ldots.$$

In the case of a finite sequence, the array terminates; in the infinite case it does not. When we use the term *sequence* by itself, it will generally be understood that it refers to an infinite sequence; sometimes it is used with reference to finite sequences, but in these cases the meaning will be clear from the context. The values of a sequence, finite or infinite, may be mathematical entities of essentially any kind, not necessarily numbers. It should also be kept in mind that the values of a sequence need not be all different.

Following long-established custom, we will often write $f_1, f_2, \ldots, f_n \ldots$ instead of $f(1), f(2), \ldots, f(n), \ldots$ respectively, for the values of a sequence. However, there are occasions, particularly when dealing with *subsequences* (cf. Definition 4.27), when the second notation is used. It is the correct notation, although at times it is cumbersome to deal with. Again, some writers find it convenient for the sake of brevity to define a sequence by giving its "general" term $\{f_n\}$ instead of defining it in the customary way for functions. For instance, if reference is made to the sequence $\{1/n\}$ with no further explanation, then the reader is expected to understand that this refers to the function f for which $\mathscr{D}(f) = P$ and $f(n) = 1/n$ for each $n \in P$. Since the reader can usually make the appropriate interpretation without difficulty, the brace notation achieves succinctness. The letter n used in the expression within the braces may be replaced by any other convenient letter, provided the new letter is not already in use in the expression, since n plays in effect the role of a dummy variable in this shorthand notation. For example, the sequences $\{1/n\}$, $\{1/m\}$, and $\{1/k\}$ are all to be understood to refer to the same sequence, namely the one whose values are 1, $\frac{1}{2}$, $\frac{1}{3}$, etc. If one writes $\{1/(n + 2k)\}$, then it is not immediately clear whether the sequence under consideration is that whose values are $1/(1 + 2k)$, $1/(2 + 2k)$, $1/(3 + 2k)$, etc., or $1/(n + 2)$, $1/(n + 4)$, $1/(n + 6)$, etc., or even something different. When one or more letters are used as parameters, then the brace notation needs additional explanatory statements. In the example just given, if k is supposed to be a parameter, then the sequence $\{1/(n + 2k)\}$ would be the one whose values are $1/(1 + 2k)$, $1/(2 + 2k)$, $1/(3 + 2k)$, etc. This would be the same as $\{1/(m + 2k)\}$, but it would not be the same as $\{1(k + 2k)\}$. The replacement of n by any letter other than k is legitimate; since k is used in the general term in a role different from that of n, the result of replacing n by k would give an undesired change of meaning. These remarks are intended to point out both the advantages and the difficulties inherent in the shorthand brace notation which relies on the reader to make correct interpretations. We shall use it when there is no possibility of confusion as to the meaning.

The device, used in many elementary textbooks, of giving only the first three or four terms of a sequence is the least desirable way of defining a sequence, and should not be used at all, since logically there is no way to determine the succeeding values regardless of how obvious the pattern of the first few may appear.

4.2 Definition.   A sequence A is said to *converge to a* or to *approach a* if and only if A is *real-valued*, that is, $\mathscr{R}(A) \subset R$, a is a real number, and for each $\epsilon > 0$ there exists $N \in P$ such that $|A_n - a| < \epsilon$ whenever $N < n \in P$.[1]

A sequence A is said to *converge* if and only if there exists $a \in R$ such that A converges to a; otherwise the sequence is said to *diverge*.

To determine from Definition 4.2 whether or not a given real-valued sequence converges, it appears to be necessary to find a real number a for which the criterion of that definition is satisfied. This can be difficult. Later we will establish *Cauchy's criterion for convergence*; it does not depend on testing possible values for a, although it has certain difficulties associated with it also.

4.3 Lemma.   If the sequence A converges to a and also converges to a', then $a = a'$.

Proof.   If we take an arbitrary positive number ϵ, then evidently $\epsilon/2 > 0$, and in accordance with Definition 4.2 there exist positive integers N and

---

[1] The notation "$N < n \in P$" is an abbreviation for "$N < n$ and $n \in P$." This form of abbreviation will be used regularly henceforth.

N' such that

(1) $$|A_n - a| < \frac{\epsilon}{2}$$ whenever $N < n \in P$ and

(2) $$|A_n - a'| < \frac{\epsilon}{2}$$ whenever $N' < n \in P.$

We let $K = N + N'$. Clearly $N < K$, $N' < K$, $K \in P$. Hence from (1) and (2) we infer that $|A_K - a| < \epsilon/2$ and $|A_K - a'| < \epsilon/2$. Thus

$$|a - a'| = |(a - A_K) + (A_K - a')|$$

$$\leq |a - A_K| + |A_K - a'| < \frac{\epsilon}{2} + \frac{\epsilon}{2} = \epsilon$$

Since ϵ is an arbitrary positive number, we may invoke § 2.2 (xvi) to conclude that $|a - a'| = 0$; hence $a = a'$ as required.

The technique employed in the proof of this lemma, in which one or more terms and their negatives are inserted so as to enable the splitting of an absolute value into two or more parts, each of which is less than a pre-arranged fraction of ϵ, so that the whole adds up to ϵ, is standard in analysis, and will be used repeatedly.

The lemma just proved assures us that if a sequence converges it approaches exactly one real number. This justifies the following definition.

4.4   Definition.   If the sequence A converges, then the real number to which it converges is called the *limit* of A. If $a \in R$ is the limit of A, then we agree to write $\lim A = a$.

For much the same reason as the brace notation was introduced to define sequences, that is, for brevity, mathematicians use certain alternatives to the notation just introduced, namely $\lim_n A_n = a$ (read: *the limit of A_n is a*) or $\lim_{n \to \infty} A_n = a$ (read: *the limit of A_n as n tends to infinity is a*). In each of these expressions n plays the role of a dummy variable and may be replaced by any other variable, provided it is not already in use in the expression for A_n. We will most frequently use the notation $\lim_n A_n = a$, and we shall avoid $\lim_{n \to \infty} A_n = a$, since the expression "$n$ tends to infinity" carries with it some unfortunate interpretations, as will be remarked shortly.

The concept of limit is one of the central ideas of mathematics. It takes many forms, although all have certain common roots.[2] Sequential limits are in some respects the simplest kind, and it is well to pause here to take a

[2] An excellent analysis of the subject may be found in the article "Partial Orderings and Moore-Smith Limits" by E. J. McShane, *Amer. Math. Assoc. Monthly*, **59**, Jan. 1952, pp. 1–11.

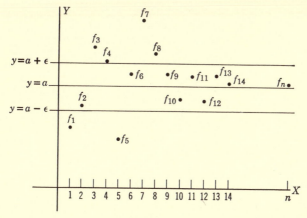

Figure 1

close look at them to see just what is involved in the definition we have given.

The limit of a real-valued sequence is just a single real number associated in a certain manner with the sequence. The definition of the limit does not require, nor does it preclude, that the values of the sequence should ever coincide with the limit. Knowledge that a sequence f has a limit a allows us to draw no conclusions as to the values assumed by the sequence.

If we imagine the value of the sequence f arranged in the natural order of its domain, namely,

$$f(1), f(2), \ldots, f(n), \ldots.$$

then our definition tells us only that as we go far out into this array, the terms tend to cluster about a with none, some, or possibly all coinciding with a. More precisely, if we are given an arbitrary positive number ϵ, and if we imagine counting off the terms $f(1)$, $f(2)$, $f(3)$, etc., eventually we will arrive at some term, say the Nth, such that *all* succeeding terms are within a numerical distance ϵ of the number a. We may represent this situation pictorially on a graph as shown above. The values f_1, f_2, f_3, etc., of the sequence f may be plotted vertically above the points 1, 2, 3, etc., respectively, on the X-axis. The lines $y = a + \epsilon$ and $y = a - \epsilon$ represent the boundaries within which the terms of the sequence must eventually lie. We may imagine that $N = 8$ in the situation depicted in our graph.

Again looking at our graph, let us imagine what happens if ϵ is decreased. Then the boundary lines $y = a + \epsilon$ and $y = a - \epsilon$ are brought closer together. It is apparent that this might necessitate going much farther out into the sequence before we arrive at a term beyond which all succeeding points representing sequence values lie between the new boundary lines. Thus it is not to be expected that if an integer N fulfills the requirements of the limit definition for a given number ϵ, it will fulfill the corresponding requirements for a smaller number ϵ'. In other words, ordinarily N is going to depend on ϵ; more specifically, the smaller ϵ may be, the larger must be N. The important thing is that no matter what positive number ϵ is given at the outset, there must exist some N satisfying the requirements of the definition, corresponding to that value of ϵ. Incidentally, it is worth noting that to prove $\lim_n f_n = a$, we need only to show the existence of a positive integer N with appropriate properties; we do not have to find the smallest positive integer with these properties. In applications we frequently find it much easier to discover a value for N that is far from being the smallest possible value; yet this is perfectly satisfactory for our purposes.

We conclude this heuristic digression with a few final remarks. Sometimes in elementary calculus books it is stated that when $\lim_n f_n = a$, the difference $f_n - a$, or perhaps $|f_n - a|$, becomes *arbitrarily small*, or *infinitesimally small, as n tends to infinity*. Along with these and similar statements, one is apt to pick up a hazy impression that somewhere "out at infinity" the values of the sequence merge in some mysterious fashion along the line $y = a$ shown on our graph. It is unfortunate that the language of these expressions tends to convey such an impression, for, unlike some intuitive pictures that are of assistance in comprehending the nature of and formalizing a concept, this one is so fuzzy it tends to obscure it. It must be borne in mind that although P is an infinite set in the formal sense of Chapter 3, each of its members is preceded by only a finite number of other members of P, and there are an infinite number of members of P greater than any given member of P, so there is no effective way to distinguish any member of P from any other in this sense; hence there is no mechanism here for "going to infinity" or "approaching infinity" in this connection. No more meaning can be attributed to the statements in italics above, nor to similar statements, than is expressed in our formal limit definition; in fact, that definition represents the culmination of the efforts of many mathematicians to endow those statements with precise meaning.

We now give a few examples of sequence with limits. We consider the sequence $\{1/n\}$. It is intuitively obvious that the terms $1, \frac{1}{2}, \frac{1}{3}$, etc., of this sequence become small numerically, and it would seem that if it has a limit at all, it must be zero. Thus we attempt to prove that $\lim_n (1/n) = 0$.

As will be seen, the Archimedean property in one form or another plays an important role in this and many other similar problems.

In accordance with Definition 4.2, we assume we are given an arbitrary positive number ϵ. We must find $N \in P$ so that whenever $N < n \in P$ the inequality

$$\left| \frac{1}{n} - 0 \right| = \frac{1}{n} < \epsilon$$

holds. By Corollary 2.11, there exists $N \in P$ such that $(1/N) < \epsilon$. Clearly, if $N < n \in P$, then

$$\left| \frac{1}{n} - 0 \right| = \frac{1}{n} < \frac{1}{N} < \epsilon,$$

whence $\lim_{n} (1/n) = 0$.

In the example just given, the inequality $(1/N) < \epsilon$ is equivalent to $(1/\epsilon) < N$. In this case we can actually find the smallest possible value for N if we wish; it is the smallest positive integer exceeding $1/\epsilon$, which is well-defined due to Corollary 2.11 and the well-ordering principle for positive integers. Thus if we should wish it, we could express the smallest or "best" value for N in terms of a direct and relatively simple functional relation in terms of ϵ.

Next, we consider the sequence $\{n^2/(n^2 - 20)\}$. On calculating a few values of this sequence, one is led to decide that if it converges, its limit is probably 1. Thus we attempt to prove that $\lim_{n} n^2/(n^2 - 20) = 1$. In accordance with Definition 4.2, we assume that ϵ is an arbitrary positive number, and we must prove the existence of $N \in P$ such that

$$(1) \qquad \left| \frac{n^2}{n^2 - 20} - 1 \right| = \left| \frac{20}{n^2 - 20} \right| < \epsilon$$

whenever $N < n \in P$.

Now if we take any positive integer $N \geq 5$, it follows that $0 < N^2 - 20 < n^2 - 20$ whenever $N < n \in P$, consequently for each such N and n,

$$(2) \qquad 0 < \frac{20}{n^2 - 20} < \frac{20}{N^2 - 20}.$$

It follows that if we can choose $N \geq 5$ so as to make the extreme right-hand member of (2) less than ϵ, then the middle term will also be less than ϵ, and since $n^2 - 20 = |n^2 - 20|$ for all $n \geq 5$, then (1) will hold. We consider the inequalities

$$\frac{20}{N^2 - 20} < \epsilon, \quad 20 < \epsilon(N^2 - 20), \quad 20 + \frac{20}{\epsilon} < N^2, \quad \sqrt{20 + \frac{20}{\epsilon}} < N.$$

The second is obtained from the first by multiplying by the positive factor $N^2 - 20$; the third is obtained from the second by transposing terms and dividing by the positive number ϵ; and the last follows from the third by taking square roots of the positive numbers concerned. It is essential that all these steps be reversible so that if N is selected to satisfy the fourth inequality, and also is not less than 5, then (2) and finally (1) will hold whenever $N < n \in P$. Thus all we need to do is select any positive integer N such that $N \geq \sqrt{20 + (20/\epsilon)}$; such integers exist by Corollary 2.8. Thus once we are given $\epsilon > 0$, we can find N so that (1) holds whenever $N < n \in P$, and therefore $\lim_{n} n^2/(n^2 - 20) = 1$.

We continue these examples by considering the sequence $\{n/(n^2 - 2)\}$. Computing a few values of the sequence, we decide that it probably converges to 0. Thus we shall try to prove that $\lim_{n} n/(n^2 - 2) = 0$. We assume that ϵ is an arbitrary positive number, and we must find $N \in P$ so that

$$(3) \qquad \left| \frac{n}{n^2 - 2} \right| = \frac{n}{|n^2 - 2|} < \epsilon$$

whenever $N < n \in P$.

If we take any positive integer $N \geq 3$, and if $N < n \in P$, then $0 < N - 2 < n - 2 < n - (2/n)$; thus for any such N and n we have

$$\left| \frac{n}{n^2 - 2} \right| = \frac{n}{n^2 - 2} = \frac{1}{n - (2/n)} < \frac{1}{n - 2} < \frac{1}{N - 2}$$

Clearly, if we can choose N so that $N \geq 3$ and also $1/(N - 2) < \epsilon$, then our desired relation will hold. But $1/(N - 2) < \epsilon$ is equivalent to $1/\epsilon < N - 2$ or $2 + (1/\epsilon) < N$, provided also $N \geq 3$. Thus all we need to do is select any positive integer $N > 2 + 1/\epsilon$, since all such integers N will be necessarily not less than 3; and such a choice for N is possible by Corollary 2.8. Then for each n satisfying $N < n \in P$, (3) will hold. Therefore $\lim_{n} n/(n^2 - 2) = 0$.

Next, let us consider the sequence f so defined that $f(n) = (-1)^n$ whenever $n \in P$. This time we might suspect that the sequence has no limit at all. This is in fact so. However, we will show only that the limit cannot be 1. To prove this we consider what is the negation of our limit definition. If "$\lim_{n} A_n = a$" is false, then there must exist some positive number ϵ', such that the inequality $|A_n - a| < \epsilon'$ does not hold for each n satisfying the relation $N < n \in P$, no matter what positive integer N we may take. That is, for some $\epsilon' > 0$ and for any $N \in P$, there exists some n such that $N < n \in P$ and $|A_n - a| \geq \epsilon'$.

In our problem the values of f alternate between $+1$ and -1. Thus if we momentarily assume that 1 is the limit, then no matter how far out into the sequence we go, there are terms still farther out differing from 1 by 2 units. Thus if we take any positive integer N, and if we consider a particular value $\epsilon' = 1$, for instance (anything not exceeding 2 would do equally well), then we may choose for n the next odd positive integer greater than N; and then evidently $|f(n) - 1| = 2 > 1 = \epsilon'$, whence, in accordance with our analysis above, it follows that 1 cannot be the limit of our sequence. Of course, following these lines, we would have to repeat this argument using an arbitrary number a in place of 1 to show that the sequence has no limit at all. Later, we will use *subsequences* to help us decide that certain sequences diverge.

We might mention one final example of a divergent sequence, namely g so defined that $g(n) = n$ if n is odd, $g(n) = 0$ if n is even. In this case it happens not to be necessary to resort to the tactics used in the previous example. For as we shall see, Lemma 4.7 says that convergent sequences must be bounded. Our sequence is not, and so we infer at once that it must diverge.

It is rather tedious having to prove limit theorems directly from the definition each time. With the aid of some general theorems we can avoid many of these difficulties. We find it convenient to introduce some terminology. If A and B are real-valued sequences and $k \in R$, then $|A|$, $A + B$, $A - B$, kA, $A \cdot B$, and A/B will be the sequences $\{|A_n|\}$, $\{A_n + B_n\}$, $\{A_n - B_n\}$, $\{kA_n\}$, $\{A_n \cdot B_n\}$, and $\{A_n/B_n\}$, respectively. All these are well-defined except possibly the last, in case $B_n = 0$ for some $n \in P$. We shall establish a special convention with respect to such quotient sequences. In case there exists a positive integer M such that $B_n \neq 0$ whenever $M \leq n \in P$, that is, if the set $S = \{n \mid n \in P \text{ and } B_n = 0\}$ is finite, then we shall arbitrarily consider $\{A_n/B_n\}$ as a real-valued sequence by agreeing to define those terms of the sequence corresponding to S as being themselves equal to zero. In this way we avoid the necessity of always being forced to specify that the sequence B in the quotient sequence A/B may not take on the value zero. The following lemma shows that our assignment of the value zero to the terms corresponding to S is of no special consequence; any other real value would serve equally well without changing whatever limit properties the quotient sequence might or might not possess.

4.5 **Lemma.** If A and B are real-valued sequences, $M \in P$, and $A_n = B_n$ whenever $M \leq n \in P$, then either both sequences fail to have a limit or they have the same limit.

Proof. Either both sequences fail to have a limit, which is one of the possible conclusions stated in the theorem, or at least one of the given sequences has a limit. We may assume in the latter case that A converges to a, say. If ϵ is an arbitrary positive number, there exists $N \in P$ such that $|A_n - a| < \epsilon$ whenever $N < n \in P$. Now let $N' = N + M$. For any $n \in P$ satisfying $N' < n$ we certainly have $N < n$ and also $M < n$, whence $|B_n - a| = |A_n - a| < \epsilon$ due to our hypotheses and the nature of N. Thus $\lim_n B_n = a$.

4.6 Definition.   A real-valued function f is *bounded* if and only if the set $\mathcal{R}(f)$ is bounded, that is, there exists a real number M such that $0 \leq M$ and $|f(x)| \leq M$ whenever $x \in \mathcal{D}(f)$. Similarly, f is *bounded above* (*below*) if and only if $\mathcal{R}(f)$ is *bounded above* (*below*), that is, there exists $Q \in R$ such that $f(x) \leq Q\,(Q \leq f(x))$ for each $x \in \mathcal{D}(f)$.

4.7 Lemma.   If the sequence A converges, then it is bounded.
Proof.   By Definition 4.2 there exists a real number a to which A converges and therefore, taking $\epsilon = 1$, there exists a positive integer N such that $|A_n - a| < \epsilon = 1$ whenever $N < n \in P$; thus

$$|A_n| = |(A_n - a) + a| \leq |A_n - a| + |a| < 1 + |a|$$

whenever $n \in (P - \hat{N})$.

We let $S = \mathcal{R}(A \mid \hat{N})$. S is a finite set by Definition 3.2, and so is bounded by Theorem 3.16; thus there exists $L \geq 0$ such that $|A_n| \leq L$ whenever $n \in \hat{N}$.

We now see that if $n \in P$, then either $n \in (P - \hat{N})$ and $|A_n| < 1 + |a|$, or else $n \in \hat{N}$ and $|A_n| \leq L$. It is therefore clear that for each $n \in P$, we must have $|A_n| < 1 + |a| + L$. Consequently A is bounded.

4.8 Theorem.   If A is a real-valued sequence and $a \in R$, then the relations $\lim_n A_n = a$, $\lim_n (A_n - a) = 0$, and $\lim_n |A_n - a| = 0$ are logically equivalent to each other.
Proof.   We see that for any $n \in P$,

$$|A_n - a| = |(A_n - a) - 0| = \big| |A_n - a| - 0 \big|.$$

Consequently, if for each $\epsilon > 0$ there exists $N \in P$ such that any one of these last expressions is less than ϵ whenever $N < n \in P$, then the same is true for the others. Reference to Definition 4.2 as applied to the sequences $\{A_n\}$, $\{A_n - a\}$, and $\{|A_n - a|\}$, respectively, completes the proof.

4.9 Theorem.   If $M \in P$, A and B are such real-valued convergent sequences that $A_n \leq B_n$ whenever $M \leq n \in P$, then $\lim_n A_n \leq \lim_n B_n$.

If, more particularly, $\lim_n A_n = \lim_n B_n$ and C is such a real-valued sequence that $A_n \leq C_n \leq B_n$ whenever $M \leq n \in P$, then $\lim_n C_n$ exists and coincides with the common limit of A and B.

Proof. We let $\lim_n A_n = a$, $\lim_n B_n = b$, and suppose, if possible, that $b < a$. We let $a - b = \epsilon' > 0$, whence $\epsilon'/2 > 0$. In accordance with Definition 4.2, there exist positive integers N' and N'' such that

(1)

$$|A_n - a| < \frac{\epsilon'}{2}, \qquad \text{that is, } a - \frac{\epsilon'}{2} < A_n < a + \frac{\epsilon'}{2}$$

$$\text{whenever } N' < n \in P,$$

$$|B_n - b| < \frac{\epsilon'}{2}, \qquad \text{that is, } b - \frac{\epsilon'}{2} < B_n < b + \frac{\epsilon'}{2}$$

$$\text{whenever } N'' < n \in P.$$

We let $N = M + N' + N''$. Since N is greater than each of M, N', and N'', we have from (1) and our hypotheses

(2) $\quad a - \dfrac{\epsilon'}{2} < A_N < a + \dfrac{\epsilon'}{2}; \quad b - \dfrac{\epsilon'}{2} < B_N < b + \dfrac{\epsilon'}{2}; \quad A_N \leq B_N.$

From the first relation of (2) we obtain $a - (\epsilon'/2) < A_N$; from the second we obtain $B_N < b + (\epsilon'/2)$. Putting these together and using the third, we get

$$a - \frac{\epsilon'}{2} < A_N \leq B_N \leq b + \frac{\epsilon'}{2},$$

from which by transposing the outer terms we obtain

$$a - b < \frac{\epsilon'}{2} + \frac{\epsilon'}{2} = \epsilon' = a - b.$$

This contradiction proves that $a \leq b$.

We now take up the second part of the theorem. We have $a = \lim_n A_n = \lim_n B_n = b$. Now we suppose that ϵ is an arbitrary positive number. Invoking Definition 4.2, there exist positive integers K' and K'' such that

(3) $\quad |A_n - a| < \epsilon$, that is, $a - \epsilon < A_n < a + \epsilon$ whenever $K' < n \in P$,

$\quad\;\; |B_n - a| < \epsilon$, that is, $a - \epsilon < B_n < a + \epsilon$ whenever $K'' < n \in P$.

From (3) and our hypotheses we see that

$$a - \epsilon < A_n \leq C_n \leq B_n < a + \epsilon$$

whenever $K' + K'' + M < n \in P$, which leads to

$$a - \epsilon < C_n < a + \epsilon, \qquad \text{that is, } |C_n - a| < \epsilon$$

for each such n. Consequently $\lim_n C_n = a$.

4.10   Definition.   If A is a sequence then we shall say that A is *essentially constant* if and only if there exist $M \in P$ and $a \in R$ such that $A_n = a$ whenever $M \le n \in P$; in case $M = 1$, we say that A is *constant*.

4.11   Lemma.   If A is an essentially constant sequence, then it converges to its essentially constant value.

Proof.   There exists $M \in P$ and $a \in R$ such that $A_n = a$ whenever $M \le n \in P$. Evidently $|A_n - a| = 0 < \epsilon$ whenever $\epsilon > 0$ and $M \le n \in P$; therefore $\lim_n A_n = a$.

4.12   Corollary.   If $M \in P$, $b \in R$, A is a convergent sequence, and $A_n \le b \, (A_n \ge b)$ whenever $M \le n \in P$, then $\lim_n A_n \le b$ $(\lim_n A_n \ge b)$.

Proof.   Define the constant sequence B so that $B_n = b$ for each $n \in P$ and apply Theorem 4.9 and Lemma 4.11.

4.13   Theorem.   If A is a convergent sequence, then so is $|A|$; furthermore, $\lim_n |A_n| = |\lim_n A_n|$.

Proof.   We let $\lim_n A_n = a$. By virtue of § 2.3 (vi) we have, for each $n \in P$,

$$0 \le \big| \, |A_n| - |a| \, \big| \le |A_n - a|.$$

We consider the sequences $\{|A_n - a|\}$, $\{|A_n| - |a|\}$, and the constant sequence $\{0\}$. By Theorem 4.8 and Lemma 4.11, the first and last of these converge to zero; hence so does $\{|A_n| - |a|\}$ by Theorem 4.9. Applying Theorem 4.8 again, we infer that $\lim_n |A_n| = |a|$.

The converse of this theorem is not true, as may be seen by considering the sequence A for which $A_n = (-1)^n$ whenever $n \in P$.

4.14   Theorem.   If A is a convergent sequence and $k \in R$, then kA is convergent; in fact $\lim_n kA_n = k(\lim_n A_n)$.

Proof.   If ϵ is an arbitrary positive number, then $\epsilon/(|k| + 1)$ is positive and there exists $N \in P$ such that

$$|A_n - a| < \frac{\epsilon}{|k| + 1}$$

whenever $N < n \in P$. Thus for each such n we have

$$|kA_n - ka| = |k| \cdot |A_n - a| < |k| \frac{\epsilon}{|k| + 1} < \epsilon,$$

whence $\lim_n kA_n = ka$.

4.15   Corollary.   If A and B are sequences, A is bounded, and B converges to zero, then $A \cdot B$ converges to zero.

Proof. We let $M \geq 0$ denote a bound for $\mathscr{R}(A)$. Then $0 \leq |A_n B_n| \leq M |B_n|$ whenever $n \in P$. By Theorems 4.14, 4.13, and Lemma 4.11, the first and last of the sequences $\{0\}$, $\{A_n B_n\}$, and $\{M |B_n|\}$ converge to zero, so by Theorems 4.9, and 4.8, $A \cdot B$ converges to zero.

4.16 Theorem. If A and B are convergent sequences, then so is $A + B$; furthermore, $\lim_n (A_n + B_n) = \lim_n A_n + \lim_n B_n$

Proof. If ϵ is an arbitrary positive number, then $\epsilon/2$ is positive and by Definition 4.2 there exist positive integers N' and N'' for which

(1)
$$|A_n - a| < \frac{\epsilon}{2} \qquad \text{whenever } N' < n \in P,$$

$$|B_n - b| < \frac{\epsilon}{2} \qquad \text{whenever } N'' < n \in P.$$

We let $N = N' + N''$ and note that $N' < N$, $N'' < N$. Consequently, if $N < n \in P$, we infer from (1) that

$$|(A_n + B_n) - (a + b)| = |(A_n - a) + (B_n - b)|$$
$$\leq |A_n - a| + |B_n - b| < \frac{\epsilon}{2} + \frac{\epsilon}{2} = \epsilon$$

whence it follows that $\lim_n (A_n + B_n) = a + b$, and the theorem is proved.

4.17 Corollary. If A and B are convergent sequences, then so is $A - B$; in fact, $\lim_n (A_n - B_n) = \lim_n A_n - \lim_n B_n$.

Proof. Apply Theorems 4.14 and 4.16 to the sequence $A - B = A + (-B)$.

4.18 Theorem. If A and B are convergent sequences, then so is $A \cdot B$; more specifically, $\lim_n A_n \cdot B_n = (\lim_n A_n) \cdot (\lim_n B_n)$.

Proof. We let $a = \lim_n A_n$, $b = \lim_n B_n$. The sequence A and the constant sequence $\{b\}$ converge and so are bounded. Also the sequences $\{B_n - b\}$ and $\{A_n - a\}$ converge to zero. Thus the sequences $\{A_n \cdot (B_n - b)\}$ and $\{b \cdot (A_n - a)\}$ converge to zero by Corollary 4.15, and so their sum, namely $\{A_n B_n - ab\}$, converges to zero by Theorem 4.16. Consequently, $\lim_n A_n B_n = ab$.

Theorems 4.16 and 4.18 can be extended inductively to cover the sums and products, respectively, of any finite number of convergent sequences. We shall not do that here, however.

4.19 Lemma. If A is a convergent sequence whose limit a is positive, then there exists $N \in P$ such that $0 < a/2 < A_n$ whenever $N < n \in P$.

Proof. Since $a/2$ is positive, there exists $N \in P$ for which

$$|A_n - a| < \frac{a}{2}, \qquad \text{that is, } a - \frac{a}{2} < A_n < a + \frac{a}{2}$$

whenever $N < n \in P$. The left-hand portion of this inequality is what we wanted.

4.20 Corollary. If A is a convergent sequence whose limit a is negative, then there exists $N \in P$ such that $A_n < a/2 < 0$ whenever $N < n \in P$.

Proof. Apply Lemma 4.19 to the sequence $-A$ with positive limit $-a$.

These last two results show that convergent sequences with non-zero limits stay at a positive distance from zero whenever we go sufficiently far out into these sequences. This property is important in connection with the following lemma.

4.21 Lemma. If A is a convergent sequence whose limit a is not zero, then $1/A$ is convergent and $\lim_n (1/A_n) = 1/a$.

Proof. Due to Lemma 4.19 and Corollary 4.20, there exists $N \in P$ such that $0 < |a|/2 < |A_n|$ whenever $N < n \in P$. Thus in accordance with our convention on quotient sequences, $1/A$ is a real-valued sequence. For each n such that $N < n \in P$ we see that

$$(1) \qquad 0 \le \left| \frac{1}{A_n} - \frac{1}{a} \right| = \frac{|a - A_n|}{|a| \cdot |A_n|} < \frac{2}{|a|^2} \cdot |A_n - a|$$

Now the sequence $\{|A_n - a|\}$ converges to zero due to our hypotheses; hence so does $\{(2/|a|^2)|A_n - a|\}$; also, so does the constant sequence $\{0\}$. Thus from the relations (1) and Theorem 4.9 we infer that $\{|(1/A_n) - (1/a)|\}$ converges to zero, whence we arrive at the desired result.

4.22 Theorem. If A and B are convergent sequences and $\lim B_n \ne 0$, then A/B is convergent and $\lim_n (A_n/B_n) = (\lim_n A_n)/(\lim_n B_n)$.

Proof. It follows that A/B coincides with $A \cdot (1/B)$. Application of Theorem 4.18 and Lemma 4.21 yields the desired result.

4.23 Definition. A sequence A is said to be *essentially non-decreasing* (*non-increasing*) if and only if there exists $N \in P$ such that $A_n \le A_{n+1}$ ($A_{n+1} \le A_n$) whenever $N \le n \in P$; it is said to be *essentially strictly increasing* (*decreasing*) if and only if $A_n < A_{n+1}$ ($A_{n+1} < A_n$) whenever $N \le n \in P$. We drop the qualifying term "essentially" in case $N = 1$.

4.24 Lemma. If A is a sequence, then A is essentially non-decreasing (non-increasing) if and only if there exists $N \in P$ and $A_i \leq A_j$ $(A_j \leq A_i)$ whenever i and j belong to P and $N \leq i < j$.

Proof. We suppose A is an essentially non-decreasing (non-increasing) sequence, let N satisfy the conditions of Definition 4.23, and define

$$T = \{n \mid n \in P \text{ and there exists } i \in P \text{ for which } N \leq i < n \text{ and } A_n < A_i\}$$
$$(T = \{n \mid n \in P \text{ and there exists } i \in P \text{ for which } N \leq i < n \text{ and } A_i < A_n\}).$$

We assume that $T \neq \varnothing$, and let j be the smallest member of T. There exists $i \in P$ such that $N \leq i < j$ and $A_j < A_i$ $(A_i < A_j)$. Clearly $j > N$ and so $j - 1 \geq N$. From our hypotheses we must have

(1) $$A_{j-1} \leq A_j < A_i \quad (A_i < A_j \leq A_{j-1}).$$

Now either $j - 1 = i$ or $i < j - 1$. If $i = j - 1$ then $A_{i+1} = A_j < A_i$ $(A_i < A_j = A_{i+1})$, contradicting the non-decreasing (non-increasing) nature of A. If $i < j - 1$ then we see from (1) that $j - 1 \in T$, contrary to the assumption that j is the smallest member of T. Thus $T = \varnothing$ and one part of our lemma is proved. The converse is obvious.

4.25 Lemma. If A is a sequence then A is essentially strictly increasing (decreasing) if and only if there exists $N \in P$ and $A_i < A_j$ $(A_j < A_i)$ whenever i and j belong to P, $N \leq i < j$.

Proof. This can be accomplished by making suitable and rather obvious replacements of $<$ and $\leq$ by $\leq$ and $<$, respectively, in the proof of the preceding theorem. The details are left to the reader.

4.26 Theorem. If A is a real-valued essentially non-decreasing (non-increasing) sequence that is bounded above (below), then A converges; in fact,

$$\lim_n A_n = \sup\left(\mathscr{R}(A)\right) \qquad \lim_n A_n = \inf\left(\mathscr{R}(A)\right).$$

Proof. Since A is bounded above (below) in accordance with Definition 4.6, it follows that $\mathscr{R}(A)$ has a supremum (infimum). We let $M = \sup \mathscr{R}(A)$ $(m = \inf \mathscr{R}(A))$. Given $\epsilon > 0$, by Theorem 2.13 there exists $y \in \mathscr{R}(A)$ such that $M - \epsilon < y \leq M$ $(m \leq y < m + \epsilon)$. Now there exists $N \in P$ such that $y = A_N$. By Lemma 4.24, if $N < n \in P$, then $A_N \leq A_n$ $(A_n \leq A_N)$ and $A_n \in \mathscr{R}(A)$; consequently

$$M - \epsilon < A_N \leq A_n \leq M \quad (m \leq A_n \leq A_N < m + \epsilon),$$

whence $|A_n - M| < \epsilon$ $(|A_n - m| < \epsilon)$ and therefore

$$\lim_n A_n = M \quad (\lim_n A_n = m).$$

4.27 **Definition.** If A is a sequence, then we shall say that B is a *subsequence of A* if and only if there exists a strictly increasing positive integer-valued sequence f such that $B = A \circ f$.

We see that if B is a subsequence of A, then in the notation of our definition we must have $\mathscr{D}(f) = P$ and $\mathscr{R}(f) \subset P = \mathscr{D}(A)$; thus by Definition 1.47 and Theorem 1.46, $\mathscr{D}(B) = P$, $\mathscr{R}(B) \subset \mathscr{R}(A)$, and B must be itself a sequence that takes on some of the values assumed by A. This fact alone would not distinguish B as a subseqence of A, however. One additional ingredient is required, furnished by the function f. We can see how this operates by writing

$$B_1 = A(f(1)), \, B_2 = A(f(2)), \ldots, B_n = A(f(n)), \ldots$$

We see that on writing B in this way in the natural order of its indices, we also write a portion (possibly all) of the sequence A. Now, due to the increasing nature of f, the indices of the corresponding terms of A, namely $f(1), f(2), \ldots, f(n), \ldots$ are also seen to be in their natural order, and without repetition. Thus a subsequence B picks out of the parent sequence A an infinite collection of its terms, and preserves their order as they appear in the parent sequence.[3]

4.28 **Lemma.** If f is a strictly increasing positive integer-valued sequence, then $f(n) \geq n$ for each $n \in P$.
Proof. We show this inductively. We let

$$Q = \{n \mid n \in P \text{ and } n \leq f(n)\}.$$

Since $f(1) \in P$, clearly $1 \leq f(1)$ and so $1 \in Q$. Next we assume that $n \in Q$. Then $n \leq f(n) \in P$, whence $n + 1 \leq f(n) + 1$. But since f is strictly increasing, $f(n) < f(n + 1) \in P$, whence $(f(n + 1) - f(n)) \in P$ and $f(n + 1) \geq f(n) + 1$. Putting together our inequalities, we obtain $n + 1 \leq f(n + 1)$; thus $n + 1 \in Q$ if $n \in Q$. Therefore $Q = P$ and our proof is complete.

4.29 **Theorem.** If A is a convergent sequence and B is a subsequence of A, then B converges to the same limit as A.
Proof. We let f be a sequence fulfilling the requirements of Definition 4.27; we set $B = A \circ f$. We let $a = \lim_{n} A_n$, assume ϵ is an arbitrary positive number, and note the existence of $N \in P$ such that

(1) $$|A_m - a| < \epsilon \quad \text{whenever} \quad N < m \in P.$$

---

[3] Professor J. L. Kelley of the University of California at Berkeley uses a weaker restriction than this, which avoids some of the difficulties inherent in the demand that order be preserved, and still possesses all the important properties of subsequences. See the article "Partial Orderings and Moore-Smith Limits" by E. J. McShane, *Amer. Math. Monthly*, **59**, Jan. 1952, pp. 1–11.

Now we take an arbitrary positive integer n such that $N < n \in P$, note from Lemma 4.28 that $N < n \leq f(n) \in P$, and use (1) to infer that

$$|B_n - a| = |A(f(n)) - a| < \epsilon.$$

Thus $\lim\limits_{n} B_n = a$ as required.

4.30 Corollary. The sequence A converges if and only if each of its subsequences converges to a common limit.
Proof. If A converges, then by the theorem just proved, each subsequence of A must converge to the limit of A itself. Conversely, suppose that every subsequence of A converges to a common limit $a \in R$. Now we consider the function f so defined on P that $f(n) = n$ whenever $n \in P$. Clearly f is a positive integer-valued strictly increasing sequence, and $A = A \circ f$. Hence A is one of its own subsequences, whence A converges to a by assumption.

This last result is very useful in determining the non-convergence of a sequence, for if one can find a single subsequence that has no limit, or two subsequences with different limits, then the entire sequence necessarily diverges.

Sometimes we have occasion to take a subsequence of a subsequence, and the following is therefore of importance.

4.31 Lemma. A subsequence of a subsequence of a given sequence A is a subsequence of A.
Proof. We consider B, a subsequence of A and C, a subsequence of B. In accordance with Definition 4.27, there exist positive integer-valued, strictly increasing sequences g and f such that $B = A \circ g$ and $C = B \circ f$. We let $h = g \circ f$. Now $P = \mathscr{D}(f)$, $\mathscr{R}(f) \subset P = \mathscr{D}(g)$ and so by Theorem 1.46, $\mathscr{D}(h) = P$ and $\mathscr{R}(h) \subset P$. If $i \in P$, $j \in P$ and $i < j$, then $f(i) < f(j)$; so $h(i) = g(f(i)) < g(f(j)) = h(j)$ since g is strictly increasing. Consequently, h is a strictly increasing positive integer-valued sequence. But also $C_n = B(f(n)) = A(g(f(n))) = A(h(n))$ for each $n \in P$, so that $C = A \circ h$. Thus C is a subsequence of A.

We come now to the *Cauchy criterion* for the convergence of a sequence. It is not always easy to apply, but when we are able to apply it, it tells us whether or not a given sequence converges without the necessity of guessing what, if anything, might be the possible limit value.

4.32 Definition. The real-valued sequence A is said to satisfy the *Cauchy criterion* or to be a *Cauchy sequence* if and only if for each $\epsilon > 0$ there exists $N \in P$ such that $|A_n - A_m| < \epsilon$ whenever $N < m, n \in P$.[4]

---

[4] The expression "$N < m$, $n \in P$" is an abbreviation we shall use henceforth for "$N < m$, $N < n$, $m \in P$, and $n \in P$."

4.33 Lemma. If A is a Cauchy sequence, then A is bounded.

Proof. We let $\epsilon = 1$, and from Definition 4.32 infer the existence of $N \in P$ such that $|A_n - A_m| < \epsilon = 1$ whenever $N < m$, $n \in P$. Thus

$$|A_n| = |(A_n - A_{N+1}) + A_{N+1}| \leq |A_n - A_{N+1}| + |A_{N+1}| < 1 + |A_{N+1}|$$

whenever $N < n \in P$, that is, whenever $n \in (P - \hat{N})$. Setting $a = A_{N+1}$, the remainder of the proof becomes identical with that of Lemma 4.7, and so is not repeated here.

4.34 Theorem. If A is a Cauchy sequence, then A converges.

Proof. We let $S = \mathscr{R}(A)$. There are two separate cases to consider.

CASE 1. S is a finite set.

We define the function Q on S so that

$$(1) \qquad\qquad Q(s) = \{n \mid n \in P \text{ and } A_n = s\}$$

whenever $s \in S$. Clearly $Q(s)$ is a subset of P for each $s \in S$. Suppose, if possible, that $Q(s)$ is a finite set for each $s \in S$. Setting $\mathscr{F} = \mathscr{R}(Q)$, keeping in mind that $\mathscr{D}(Q) = S$ is a finite set, and applying Theorem 3.15, it follows that $\mathscr{F}$ must be a finite family of finite subsets of P, whence $\bigcup \mathscr{F}$ must be a finite subset of P by virtue of Theorem 3.9 and so must have a largest member $K \in P$ by Theorem 3.20. However, $A_{K+1} \in \mathscr{R}(A) = S$, and there exists $s' \in S$ for which $A_{K+1} = s'$, whence $K + 1 \in Q(s')$, $Q(s') \in \mathscr{F}$, and so $K + 1 \in \bigcup \mathscr{F}$, in contradiction to the fact that K is the largest member of $\bigcup \mathscr{F}$. Thus we conclude that there exists $s_0 \in S$ for which $Q(s_0)$ is an infinite set, and we assert that $s_0 = \lim_n A_n$.

To prove this we take an arbitrary positive number ϵ, and from the Cauchy criterion ascertain the existence of $N \in P$ such that

$$(2) \qquad\qquad |A_n - A_m| < \epsilon \quad \text{whenever} \quad N < m, n \in P.$$

Since $Q(s_0)$ is an infinite set, it cannot be a subset of the finite set $\hat{N}$; hence there exists $k \in Q(s_0) \subset P$ such that $k \notin \hat{N}$, that is, $N < k \in P$. From (1) and (2) we see that $A_k = s_0$ and $|A_n - s_0| = |A_n - A_k| < \epsilon$ whenever $N < n \in P$; consequently $s_0 = \lim_n A_n$.

CASE 2. S is an infinite set.

By Lemma 4.33 S is bounded, and so by Theorem 3.25 S has a cluster point y. We will show that y is the limit of the sequence A.[5] We suppose

---

[5] The Bolzano-Weierstrass theorem guarantees the existence of at least one cluster point, but there may be many. When we prove that the limit of the sequence A coincides with one of them, as we shall do, then since limits of sequences are unique, we will have shown in effect that there is only one cluster point in this case.

ϵ is an arbitrary positive number. Then the set

$$T = S \cap \left\{ x \,\middle|\, |y - x| < \frac{\epsilon}{2} \right\}$$

is an infinite subset of S due to the fact that y is a cluster point of S. Also, there exists $M \in P$ such that

(3) $$|A_m - A_n| < \frac{\epsilon}{2} \qquad \text{whenever } M < m, n \in P,$$

due to the Cauchy criterion. Since the set $S' = \mathscr{R}(A \mid \hat{M})$ is a finite subset of S, then T cannot be contained in S', whence there exists $x \in (T - S')$. Now there exists $p \in P$ such that $x = A_p$; $M < p$ since $A_p \notin S'$, and $|y - A_p| < \epsilon/2$ due to the membership of $A_p = x$ in T. From these facts and (3), we see that

$$|A_n - y| \le |A_n - A_p| + |A_p - y| < \frac{\epsilon}{2} + \frac{\epsilon}{2} = \epsilon$$

whenever $M < n \in P$, whence $\lim_n A_n = y$.
This completes our proof.

4.35 Theorem. If A converges, then it is a Cauchy sequence.
Proof. If $\lim_n A_n = a$ and ϵ is an arbitrary positive number, then there exists $N \in P$ such that

(1) $$|A_k - a| < \frac{\epsilon}{2} \qquad \text{whenever } N < k \in P.$$

We take arbitrary positive integers m and n, both greater than N. By virtue of (1) we have

$$|A_n - A_m| = |(A_n - a)| + |(a - A_m)|$$

$$\le |A_n - a| + |a - A_m| < \frac{\epsilon}{2} + \frac{\epsilon}{2} = \epsilon$$

and the proof is complete.

We consolidate the last two results in the following statement for convenience.

4.36 Corollary. A necessary and sufficient condition that the real-valued sequence A converge is that it be a Cauchy sequence.

Exercises

§§ 4.1 to 4.4

1. In the case of the sequence $\{1/n\}$ considered in the text, if $\epsilon = 0.0001$, what choice could you make for N? In the example $\{n^2/(n^2 - 20)\}$, what choice

could you make for N if $\epsilon = 0.1$? If $\epsilon = 0.004$? In the example $\{n/(n^2 - 2)\}$ what choice could you make for N if $\epsilon = 21.5$? If $\epsilon = 0.8$? If $\epsilon = 0.0002$?

In each of the following problems, prove the statement from the definition of limits.

2. $\lim_n 3/(2n + 1) = 0$

3. $\lim_n 1/(2n^2 - 5) = 0.$

4. $\lim_n n^2/(n^2 + 7) = 1.$

5. $\lim_n 10n/(n^2 - 50) = 0.$

6. $\lim_n (-1)^n/(3n - 1) = 0.$

7. $\lim_n 2n/(3n - 1) = \frac{2}{3}.$

8. $\lim_n n^3/(2n^3 - 100) = \frac{1}{2}.$

9. If $x > 1$ then $\lim_n 1/x^n = 0$. (*Hint:* See Problem 42 in the Exercises at the end of Chapter 2.)

10. $\lim_n (1/n!) = 0.$

11. Let f be a sequence such that

$$f_n = 1/n \text{ if } n \text{ is an odd positive integer}$$

$$f_n = n/(n^2 - 2) \text{ if } n \text{ is an even positive integer.}$$

Show that $\lim_n f_n = 0$. (*Hint:* From the examples in the text we know that $1/n < \epsilon$ whenever $1/\epsilon < N'$ and $N' < n \in P$; also $n/(n^2 - 2) < \epsilon$ whenever $2 + (1/\epsilon) < N''$ and $N'' < n \in P$. Thus for odd values of n, we need to select $N > 1/\epsilon$ and for even values of n, we need to select $N > 2 + (1/\epsilon)$. Evidently if we take N to be a number larger than both $1/\epsilon$ and $2 + (1/\epsilon)$, it will do whether n is even or odd. In this case it is apparent that we may take $N > 2 + (1/\epsilon)$. Incidentally, it happens to be true, although it is not a legitimate assumption at this point, that the limit of f may be ascertained by considering separately each of the sequences that go to make up f.)

12. Let f be such a sequence that

$$f = n^2/(n^2 - 20) \text{ when } n \text{ is an odd positive integer,}$$

$$f = n/(n + 1) \text{ when } n \text{ is an even positive integer.}$$

Prove that $\lim_n f_n = 1$. What value for N would you choose for an arbitrary positive number ϵ?

13. Prove that $\lim_n \dfrac{\sqrt{n + 1}}{2n^2 - 25} = 0$. *Hint:* Multiply the numerator and denominator of the fraction by $\sqrt{n + 1}$. You may assume that $\sqrt{n + 1} > 1$ for each $n \in P$.)

14. Show that $\sqrt{m} > \sqrt{n}$ whenever $m, n \in P$, $m > n$; prove that $\mathscr{R}(\{\sqrt{n}\})$ is unbounded; then use these facts to prove that $\lim_n (1/\sqrt{n}) = 0.$

15. If $k \in P$, show that $\sqrt[k]{m} > \sqrt[k]{n}$ whenever $m, n, \in P$, $m > n$; prove that $\mathscr{R}(\{\sqrt[k]{n}\})$ is unbounded, and use this to prove that $\lim_n \dfrac{1}{\sqrt[k]{n}} = 0$.

16. Prove that $\lim_n \dfrac{1}{\sqrt[3]{2n - 9}} = 0$.

17. Prove that $\lim_n \sqrt{\dfrac{n}{n + 1}} = 1$. *Hint:* Note that if $n \in P$, then

$$\left| \sqrt{\frac{n}{n + 1}} - 1 \right| = \frac{\left| \sqrt{\dfrac{n}{n + 1}} - 1 \right| \cdot \left| \sqrt{\dfrac{n}{n + 1}} + 1 \right|}{\left| \sqrt{\dfrac{n}{n + 1}} + 1 \right|} < \frac{1}{n + 1}.$$

18. Prove that $\lim_n \sqrt{\dfrac{9n^2}{n^2 + 3}} = 3$.

19. Prove that $\lim_n \sqrt{\dfrac{16n^2}{25n^2 - 7}} = \dfrac{4}{5}$.

20. Prove that if $x \in R$ then $\lim_n (x^n/n!) = 0$. (*Hint:* If $|x| \leq 1$, then $|x|^n \leq 1$ holds for each $n \in P$, and the desired conclusion can be made to depend upon Problem 10 above. If $|x| > 1$, then there exists a positive integer M such that $|x|/(M + 1) = r < 1$ (why?). Hence for any n satisfying $M + 1 \leq n \in P$ we can write

$$\frac{|x|^n}{n!} = \left(\frac{|x|^M}{M!} \right) \cdot \left(\frac{|x|}{M + 1} \cdot \frac{|x|}{M + 2} \cdots \frac{|x|}{n} \right) \leq \frac{|x|^M}{M!} r^{n - M} = \frac{|x|^M}{M! \, r^M} \cdot r^n$$

Now complete the problem using a technique based on that of Problem 9 above.

21. Use the technique discussed in the text to show that the example $f(n) = (-1)^n$ for each $n \in P$ therein considered has no limit whatever.

§§ 4.5 to 4.26

Evaluate the limits indicated in the first problems, justifying your evaluations either by proving from the definition, or by applying theorems on limits. You may assume that $\lim_n (1/n) = 0$.

22. If $k \in P$, show that $\lim_n 1/n^k = 0$. Use induction.

23. Evaluate $\lim_n \dfrac{-2n^4 + n^2}{7n^4 - 10}$. $\left(\text{*Hint:* Note that } \dfrac{-2n^4 + n^2}{7n^4 - 10} = \dfrac{-2 + (1/n^2)}{7 - (10/n^4)} \text{ for each } n \in P. \right)$

24. Evaluate $\lim_n \dfrac{-n^2 + 2n - 5}{3n^3 - 10n + 1}$.

25. Consider the sequence f defined by

$$f(n) = \frac{a_0 n^k + a_1 n^{k-1} + \ldots \ldots + a_k}{b_0 n^p + b_1 n^{p-1} + \ldots \ldots + b_p}$$

for each $n \in P$, where $b_0 \neq 0$, $k \in P$, and $p \in P$. Show that $\lim_n f(n) = 0$ if $k < p$, $\lim_n f(n) = a_0/b_0$ if $k = p$.

26. Prove that $\lim_n \dfrac{n}{h^n} = 0$ if $|h| > 1$. $\left(\text{Hint: Let } q = |h|; \text{ then } q = 1 + h' \right.$ where $h' > 0$. Show inductively that for $n > 2$, $q^n > \dfrac{n(n-1)}{2} h'^2$. Thus $|n/h^n| \leq 2/(n-1)h'^2$ for each n, $2 < n \in P$. $\Big)$

27. Evaluate $\lim_n \dfrac{n!}{n^n}$. (*Hint:* Look at Problem 20 above.)

28. Let A and B be such sequences that $\lim_n B_n = 0$ and there exists a positive integer N such that $0 \leq |A_n| \leq |B_n|$ whenever $N \leq n \in P$. Prove that $\lim_n A_n = 0$.

29. If A, B, and C are convergent sequences, prove that

$$\lim_n (A_n + B_n + C_n) = \lim_n A_n + \lim_n B_n + \lim_n C_n \text{ and}$$

$$\lim_n (A_n \cdot B_n \cdot C_n) = (\lim_n A_n) \cdot (\lim_n B_n) \cdot (\lim_n C_n).$$

30. If C is a finite sequence of convergent infinite sequences, that is, there exists $k \in P$ such that C_i is a convergent sequence for each $i \in \hat{k}$, then

$$\lim_n (C_1(n) + C_2(n) + \ldots + C_k(n))$$
$$= \lim_n C_1(n) + \lim_n C_2(n) + \ldots + \lim_n C_k(n)$$

and $\lim_n (C_1(n) \cdot C_2(n) \ldots C_k(n))$
$$= (\lim_n C_1(n)) \cdot (\lim_n C_2(n)) \ldots (\lim_n C_k(n)).$$

You may assume that the properties of finite summation and products are known.

31. Let $x \in R$, $|x| < 1$, and assume the sequence f is so defined that

$$f(n) = 1 + x + \ldots + x^n = \frac{1 - x^{n+1}}{1 - x}$$

for each $n \in P$. Prove that $\lim_n f(n) = \dfrac{1}{1 - x}$

32. Suppose that two convergent sequences A and B are so related that $A_n < B_n$ for each $n \in P$. Show by suitable examples that it is not necessarily true that $\lim_n A_n < \lim_n B_n$.

33. Given the sequence f defined so that $f(n) = 2 + \dfrac{1}{2!} + \dfrac{1}{3!} + \ldots + \dfrac{1}{n!}$ for each $n \in P$. Prove inductively that $f(n) \leq 2 + \dfrac{1}{2} + \dfrac{1}{4} + \ldots + \dfrac{1}{2^n}$ for each $n \in P$, and use the result to Problem 31 to infer that $f(n) < 3$ for each $n \in P$. Now prove that f converges. The limit of this sequence is $e = 2.71828 \ldots$.

34. Carry out the details of the proof of Lemma 4.25.

§§ 4.27 to 4.36

Use any methods available to you to determine the convergence or divergence of the sequences in the first six problems.

35. $f(n) = \dfrac{n}{n+1}$ if n is even; $f(n) = \dfrac{2n - 101}{2n}$ if n is odd.

36. $f(n) = -2^n$ if n is even; $f(n) = n^2$ if n is odd.

37. $f(n) = \dfrac{2n + 1}{n}$ if n is divisible by 3;

$f(n) = \dfrac{2^{n+1} - 1001}{2^n + 1}$ if $n = 3k - 1$ for some $k \in P$;

$f(n) = 2$ if $n = 3k - 2$ for some $k \in P$.

38. f is a strictly decreasing sequence such that $f(n + 1) \geq f(n) - (\tfrac{2}{3})^n$ for each $n \in P$.

39. $f(n) = \dfrac{10^6}{n}$ if n is even; $f(n) = 0$ if n is odd.

40. $f(n) = (\tfrac{3}{4})^n + 1$ if n is even; $f(n) = 1 - \dfrac{1}{n^2}$ if n is odd.

41. Use the Cauchy criterion to prove Theorem 4.26.

In the following five problems determine whether or not A converges. If it does converge, can you tell what its limit must be? Either prove that the stated condition forces A to be a constant sequence, or else show by a suitable example for A that it does not.

42. $|A_m - A_n| < 1/(m + n)$ for each $m, n \in P$.
43. $|A_m - A_n| < (m + n)/(m^2 + n^2)$ for each $m, n \in P$.
44. $|A_m - A_n| < mn/(m^2 + n^2)$ for each $m, n \in P$.
45. $|A_m - A_n| < m^2/(m + n)$ for each $m, n \in P$.
46. $|A_m - A_n| < mn/(m + n)$ for each $m, n \in P$.
47. Show that if Y is a subsequence of the sequence X and if f is a function whose domain includes $\mathcal{R}(X)$, then $\{f(Y_n)\}$ is a subsequence of $\{f(X_n)\}$.

5

Sequential Limit Theory in
the Extended Real Number System

In the real number system there is neither a largest nor a smallest number; thus some sets of real numbers have no upper bound at all and some have no lower bound, to say nothing of having a least upper bound or a greatest lower bound, respectively. It would be very convenient if every set of real numbers were to have some upper bound and some lower bound, and even better if each such set were to have a supremum and an infimum. We know already that every set bounded above must have a supremum and every set bounded below has an infimum. It would be nice if we could remove the qualification of boundedness in these statements. Similarly, every non-decreasing sequence that is bounded above has a limit, as does every non-increasing sequence that is bounded below. Again it would be nice if we could remove the condition of boundedness. One might thus be led to question what could be done to the real number system in order to achieve one or more of these desirable goals.

Simple boundedness of each set of real numbers could be assured by adjoining to the real number system two new elements, one of which would be the "largest" member of the so expanded system and the other the "smallest" member of it. However, such a procedure must be undertaken with considerable caution. For one thing, adding elements at the top and bottom of our scale of order might conceivably introduce hidden contradictions into the system that would destroy its usefulness. Even though it turns out that this does not happen, these exceptional elements, introduced for the purpose of eliminating exceptions in our existing system, might well lead to new and unwanted exceptions in the extended system of which they would be a part, perhaps nullifying the advantages gained by their introduction. In this connection there arises the difficult question of extending the operations of addition and multiplication to such new elements. All the ordinary elements of the system enjoy certain properties with respect to addition and multiplication and their inverses. However, these operations were defined rather naturally on the natural numbers and then extended rather naturally as that number system was expanded to the full real number system. It is not at all easy to see how to fit into this complex scheme two new elements that have simply been invented out of thin air.

5.1 The Extended Real Number System.

Aware of the possible hazards, we nevertheless agree to introduce into our system the new elements $+\infty$ and $-\infty$ (read: *plus infinity* and *minus infinity*, respectively), with the understanding that $+\infty \neq -\infty$, and that $+\infty$ and $-\infty$ are different from all members of R. As we will see shortly, $+\infty$ will be the largest member of our new system, and $-\infty$ will be its smallest member. We shall name this new system the *extended real number system*, abbreviated *Re*. Thus $Re = R \cup \langle +\infty \rangle \cup \langle -\infty \rangle$. We shall continue to use the expression "real number" to refer to members of R; however, on occasions when we deem it desirable to emphasize the fact that we are referring to members of R only we may add the prefix "finite."

The word "infinity" used in connection with $+\infty$ and $-\infty$ bears no relation to the term "infinite" as defined in Chapter 4 and has no real significance. It may be interpreted only as suggesting something farther out to the right or to the left, respectively, than any real number on the real axis.

We will extend our order relation by adjoining to it the following ordered pairs: $(x, +\infty)$ whenever $+\infty \neq x \in Re$, and $(-\infty, x)$ whenever $-\infty \neq x \in Re$. Thus we have established the relations $-\infty < x$, $-\infty < +\infty$, and $x < +\infty$ whenever $x \in R$.[1]

We would like to know if the trichotomy property of (i), § 2,2, is still valid. To this end we take arbitrary elements x and y of *Re*. Exactly one of nine possible situations must occur: $x = -\infty$, $y = -\infty$; $x = -\infty$, $y \in R$; $x = -\infty$, $y = +\infty$; $x \in R$, $y = -\infty$; $x \in R$, $y \in R$; $x \in R$, $y = +\infty$; $x = +\infty$, $y = -\infty$; $x = +\infty$, $y \in R$; $x = +\infty$, $y = +\infty$. By an examination of each of these cases it may be seen that exactly one of the relations $x = y$, $x < y$, and $y < x$ holds, so the trichotomy is preserved.

We wish to determine whether the transitive law still holds. To this end we take arbitrary members x, y, and z of *Re*, such that $x < y$ and $y < z$. It is easily checked that exactly one of four possible situations must occur: $x = -\infty$, $y \in R$, $z \in R$; $x = -\infty$, $y \in R$, $z = +\infty$; $x \in R$, $y \in R$, $z \in R$; $x \in R$, $y \in R$, $z = +\infty$. It is now easily verified that $x < z$ holds in each of these cases, consequently transitivity is preserved.

The remaining properties on order in § 2.2 relate to addition, subtraction, multiplication, and division. Since we are not going to extend these operations just now to $+\infty$ and $-\infty$, the extension of our order relation has no effect on them.

---

[1] We should give the extended order relation a new name, but it is convenient to continue using the same name, and we shall do so.

We naturally extend the relation $\leq$ so that $x \leq y$ if and only if $x \in Re$, $y \in Re$ and $x < y$ or $x = y$. Similarly, we extend $>$ and $\geq$ in such a way that $x > y$ if and only if $x \in Re$, $y \in Re$ and $y < x$; $x \geq y$ if and only if $x \in Re$, $y \in Re$ and $y > x$ or $y = x$.

It is now apparent that every non-empty subset of Re has an upper bound, namely $+\infty$, and a lower bound, namely $-\infty$. As we will show later, even the empty set has upper and lower bounds. However, we will still reserve the expressions "bounded above," "bounded below," and "bounded" to refer to those subsets of Re with ordinary real numbers as upper, lower, and general bounds, respectively.

It happens that we obtain additional bonuses as a result of our extension. One of these appears in the following theorem.

5.2   Theorem.   Every subset of Re has a supremum and an infimum belonging to Re.

Proof.   We take an arbitrary set $S \subset Re$, and bear in mind that M is the supremum of S if and only if $M \in Re$, M is an upper bound of S, and if $x < M$ then x is not an upper bound of S; similarly, m is the infimum of S if and only if $m \in Re$, m is a lower bound of S, and if $m < x$ then x is not a lower bound of S.

We shall prove only that S must have a supremum. This is done by considering four separate cases. The proof that S must have an infimum can be accomplished under four corresponding cases obtained by replacing the word "upper" by "lower" in the statment of cases 1, 2, and 3, and then making appropriate obvious modifications in the proofs of all four cases below.

CASE 1. $S \neq \varnothing$; S has no real upper bounds.

Clearly $s \leq +\infty$ whenever $s \in S$; therefore $+\infty$ is an upper bound of S. If $+\infty \neq x \in Re$, then either $x \in R$ or $x = -\infty$. By assumption, if $x \in R$ then x is not an upper bound of S. If $x = -\infty$ were an upper bound of S we would have $s \leq x < 0$ for each $s \in S$, making 0 an upper bound of S contrary to the assumption of Case 1. Hence $+\infty$ is the supremum of S.

CASE 2. $S \neq \varnothing$; there exists a real upper bound u of S and a real number $x \in S$.

Since $s \leq u < +\infty$ for each $s \in S$, then $+\infty \notin S$ and so $S = S - \langle +\infty \rangle$. We let $S' = S - \langle -\infty \rangle = S - (\langle +\infty \rangle \cup \langle -\infty \rangle)$. Evidently S' is a non-empty subset of R. Also u is an upper bound of S, and so also of S'.[2] Therefore S' has a supremum $M \in R$. Now for any element $s \in S$ we have $s \in S'$ or $s = -\infty$, whence in either case $s \leq M$ and M is an upper bound of S. On the other hand, if $M > y \in R$, then either $y \in R$

---

[2] See Problem 66 of the Exercises at the end of Chapter 2.

or $y = -\infty$. If $y \in R$, then y certainly can not be an upper bound of S', and so also is not an upper bound of S. If $y = -\infty$ then $y < x \in S$ and again y is not an upper bound of S. Thus M is the supremum of S.

CASE 3. $S \neq \varnothing$; there exists a real upper bound u of S; s has no real members.

Evidently $S \subset (\langle +\infty \rangle \cup \langle -\infty \rangle)$, and $s \leq u < +\infty$ whenever $s \in S$. Thus since $S \neq \varnothing$ it follows that $S = \langle -\infty \rangle$. Then $-\infty$ is an upper bound of S and since no members of Re are less than $-\infty$, we see that $-\infty$ is the supremum of S.

CASE 4. $S = \varnothing$.

We have deferred consideration of this case until the last because it is a rather interesting example of a vacuously true argument, and the conclusion is somewhat startling until one becomes accustomed to it.

In order that $M \in Re$ shall be an upper bound of a set $T \subset Re$, the statement "for all x, $x \in T$ implies $x \leq M$" must be true. Now if $T = \varnothing$, the first part of the statement, namely, $x \in \varnothing$, is false, making the entire statement true regardless of what M happens to be. Consequently, any member of Re is an upper bound of $\varnothing$; in particular $-\infty$ is an upper bound of $\varnothing$, and since $-\infty$ is the least member of Re, $-\infty$ is the least upper bound of $\varnothing$. Analogously one can show that $+\infty$ is the greatest lower bound of $\varnothing$.

Given any set $S \subset Re$, we see that insofar as its upper bound properties are concerned, it must fall into exactly one of the four cases just considered, and insofar as its lower bound properties are concerned, it must also fall into just one of the corresponding cases. Thus in all cases, S has a supremum and an infimum in Re.

The result noted in Case 4, namely that $-\infty = \sup \varnothing < \inf \varnothing = +\infty$, is peculiar to the null set, as we now observe.

5.3 Corollary. If $\varnothing \neq S \subset Re$, then $\inf S \leq x \leq \sup S$ whenever $x \in S$, and $\inf S \leq \sup S$.

Proof. Since $S \subset Re$, Theorem 5.2 guarantees us the existence of $\inf S$ and $\sup S$ in Re, and by definition of these things, if $x \in S$, then surely $\inf S \leq x$ and $x \leq \sup S$, which in turn implies $\inf S \leq \sup S$. However, the statement just written is in the form of an implication, and the conclusion "$\inf S \leq \sup S$" can be inferred from it only if the antecedent "$x \in S$" is not false for all x. Since $S \neq \varnothing$, there exists y such that "$y \in S$" is true, and therefore our desired conclusion is true.

5.4 Theorem. If A and B are subsets of Re and $A \subset B$, then $\inf B \leq \inf A$ and $\sup A \leq \sup B$.

Proof. For all x, if $x \in A$, then $x \in B$ and $\inf B \leq x$, $x \leq \sup B$ by Corollary 5.3. Hence $\inf B$ is a lower bound for A and $\sup B$ is an upper bound for A. Since inf A is the greatest lower bound for A and sup A is the least upper bound for A, then clearly inf $B \le$ inf A and sup $A \le$ sup B.

This theorem is true, in particular, if A and B are bounded non-empty subsets of R and their infima and suprema are, consequently, real numbers.

The following is the analogue of Theorem 2.13. It has to be recast slightly due to the peculiarities of Re. It is easy to see that if sup A (inf A) and K are finite real numbers, then it reduces essentially to Theorem 2.13.

5.5   Theorem.   If S is a subset of Re, $u \in Re$, and $u <$ sup S (inf $S < u$), then there exists $x \in S$ such that $u < x \le$ sup S (inf $S \le x < u$).
Proof.   If $u <$ sup S (inf $S < u$), then u is not an upper (lower) bound of S; consequently there exists $x \in S$ such that $u < x$ and $x \le$ sup S ($x < u$ and inf $S \le x$) due to the very definitions of the terms involved.

We introduced the concepts of cluster point for sets of real numbers in Chapter 3 (cf. Definition 3.22) and convergence of real-valued sequences in Chapter 4 (cf. Definition 4.2). We would like to extend these concepts somewhat.

5.6   Definition.   The set $I \subset Re$ is said to be a *neighborhood* of $+\infty(-\infty)$ if and only if there exists a finite positive[3] real number M such that

$$I = \{x \mid x > M\} \, (I = \{x \mid x < -M\}).$$

5.7   Definition.   We shall say that $+\infty(-\infty)$ is a *cluster point* of the set $S \subset Re$ if and only if $S \cap \{x \mid x > M\}(S \cap \{x \mid x < -M\})$ is an infinite set for each finite positive real number M.

This condition is sometimes expressed by saying that *every neighborhood of $+\infty$ ($-\infty$) contains infinitely many members of S*.

5.8   Theorem.   If S is a subset of Re, then $+\infty(-\infty)$ is a cluster point of S if and only if $S \cap \{x \mid x > M\} - \langle +\infty \rangle$ $(S \cap \{x \mid x < -M\} - \langle -\infty \rangle)$ is non-empty for each positive finite real number M.
Proof.   If $+\infty$ $(-\infty)$ is a cluster point of S, then by Definition 5.7, the sets in question are obviously infinite for each such M and so are non-empty.
If there exists a finite positive real number M' such that the set

$$A = S \cap \{x \mid x > M'\} - \langle +\infty \rangle \quad (A = S \cap \{x \mid x < -M'\} - \langle -\infty \rangle)$$

is finite, then since $A \subset R$, A is bounded above (below) by a finite real number M'' ($-M''$), where $0 < M''$, due to Theorem 3.16. It is easily seen that

$$S \cap \{x \mid x > M''\} - \langle +\infty \rangle = A \cap \{x \mid x > M''\} = \varnothing$$
$$(S \cap \{x \mid x < -M''\} - \langle -\infty \rangle = A \cap \{x \mid x < -M''\} = \varnothing).$$

---

[3] The condition of positiveness is entirely dispensable; however, it simplifies future considerations to impose this restriction.

Hence we see that if $+\infty$ $(-\infty)$ is not a cluster point of S, then there exists a neighborhood of $+\infty$ $(-\infty)$ containing no members of S, except possibly $+\infty$ $(-\infty)$. The proof is complete.

Although the sets comprising neighborhoods of $+\infty$ and $-\infty$ are different in nature from those comprising neighborhoods of finite numbers, we observe from Definitions 3.22 and 5.7 and Theorems 3.23 and 5.8 that cluster points have the same properties expressed in the language of neighborhoods whether or not such points are finite.

Examination of Definition 4.2 reveals that convergence of real-valued sequences can be expressed in terms of neighborhoods, and this fact is the basis for the following extension of our concept of sequential limits.

5.9 Definition. For any sequence A whose values are members of Re, we shall agree that $a \in Re$ is a *limit of A* if and only if for any neighborhood I of a, there exists a positive integer N such that $A_n \in I$ whenever $N < n \in P$. If A has a finite limit, then A is said to *converge*; otherwise it is said to *diverge*.

We observe that if $a = +\infty$ $(a = -\infty)$, then this definition can be expressed as follows: for each finite real number $M > 0$, there exists a positive integer N such that $A_n > M$ $(A_n < -M)$ whenever $N < n \in P$. Also, if $a \in R$ and $\mathscr{R}(A) \subset R$, then it agrees with Definition 4.2.

5.10 Extension of Addition and Multiplication. As it turns out, it is impossible to extend the operations of addition and multiplication to $+\infty$ and $-\infty$ and still maintain the validity of the rules of algebra expressed in § 2.1 as well as some of the properties of order that involve these operations. The proof of this is outside the scope of this course and will not be pursued here. Nevertheless, it is desirable to be able to perform some calculations with these infinite elements. The motivation for this is to be found in some consequences of our definition of limits for sequences taking values in Re. Consequently, we will agree to the following limited extension of the operations of addition, multiplication, and their inverses.

$$x + (+\infty) = (+\infty) + x = x - (-\infty) = +\infty \text{ whenever } -\infty \neq x \in Re;$$

$$x + (-\infty) = (-\infty) + x = x - (+\infty) = -\infty \text{ whenever } +\infty \neq x \in Re;$$

$$x \cdot (+\infty) = (+\infty) \cdot x = +\infty \text{ whenever } 0 < x \in Re;$$

$$x \cdot (-\infty) = (-\infty) \cdot x = -\infty \text{ whenever } 0 < x \in Re;$$

$$x \cdot (+\infty) = (+\infty) \cdot x = -\infty \text{ whenever } 0 > x \in Re;$$

$$x \cdot (-\infty) = (-\infty) \cdot x = +\infty \text{ whenever } 0 > x \in Re;$$

$$x/(+\infty) = x/(-\infty) = 0 \text{ whenever } x \in R;$$

$$-(+\infty) = -\infty; \ -(-\infty) = +\infty; \ |+\infty| = |-\infty| = +\infty.$$

If A and B are sequences taking values in Re, then we may wish to form the sequences $|A|$, $A + B$, $A - B$, kA, $A \cdot B$, and A/B as we did in Chapter 4. There we found it convenient to adopt a special convention with respect to A/B in case there exists $n \in P$ such that $B_n = 0$. Because our operations of addition, multiplication, and their inverses are not defined for all ordered pairs in $Re \times Re$, we now find it desirable to adopt the same sort of convention for all the sequences just named except $|A|$. Thus we agree that if at most a finite number of terms in any of these sequences are not defined, that is, if there exists $N \in P$ such that $A_n \in Re$ for all n satisfying $N \le n \in P$, then we shall still consider such sequences as having values in Re. We shall arbitrarily define the values of the undefined terms to be zero. As it will appear from Lemma 5.12, the assignment of the value zero is of no special consequence; any other value would do equally well.

5.11   Theorem.   If A is a sequence, $\mathscr{R}(A) \subseteq Re$, a and a' are limits of A, then $a = a'$.

Proof.   If $a \in R$ then there exists $Q \in P$ such that $|A_n - a| < 1$ whenever $Q < n \in P$. Due to the rules adopted in § 5.10, it follows that $A_n \in R$ whenever $Q < n \in P$; hence $|A_n| < |a| + 1$ whenever $Q < n \in P$. If $a' = +\infty \ (-\infty)$ then there exists $Q' \in P$ such that $A_n > |a| + 1 \ (A_n < -(|a| + 1))$ whenever $Q' < n \in P$. Thus $|A_n| > |a| + 1$ for each such n, and so $|a| + 1 < |A_{Q+Q'}| < |a| + 1$. Therefore we cannot have a finite and a' infinite. For this reason we are justified in restricting our considerations to the following separate cases.

CASE 1. a and a' are both finite.

From the observations just made it follows that there exists $Q \in P$ such that $A_n \in R$ whenever $Q < n \in P$. We may now take over the proof of Lemma 4.3 verbatim to infer that $a = a'$.

CASE 2. a and a' are both infinite.

If it were possible that $a = +\infty$ and $a' = -\infty$, then there must exist $Q \in P$ such that $A_n > 1$ whenever $Q < n \in P$, and $Q' \in P$ such that $A_n < -1$ whenever $Q' < n \in P$. This would require that $A_{Q+Q'} < -1$ and $A_{Q+Q'} > 1$, a contradiction. Similarly it would be impossible to have $a = -\infty$ and $a' = +\infty$. Thus we conclude that $a = a' = +\infty$ or $a = a' = -\infty$.

This completes the proof, and we now know that sequential limits, if they exist, are unique. This furnishes us with the justification for using the notation introduced in Definition 4.4 to denote limits of sequences taking values in Re.

5.12   Lemma.   If A and B are sequences with values in Re, $N \in P$, and $A_n = B_n$ whenever $N \le n \in P$, then either A and B have a common limit or none at all.

Proof. If A has a limit $a \in Re$, then given any neighborhood I of a, there exists $Q \in P$ such that $A_n \in I$ whenever $Q < n \in P$. But then $B_n \in I$ whenever $Q + N < n \in P$; thus $\lim_n B_n = a$. It follows that if one of the given sequences has a limit, the other has the same limit. Otherwise neither has a limit.

We will now give two examples of sequences with infinite limits. We start by showing that $\lim_n n^2/(n + 1) = +\infty$. We suppose that M is an arbitrary finite positive number. Now

$$\frac{n^2}{n + 1} = n\left(\frac{1}{1 + (1/n)}\right) \geq \frac{n}{2}$$

for each $n \in P$. Thus if we select N to be any positive integer exceeding $2M$ we must have

$$\frac{n^2}{n + 1} \geq \frac{n}{2} > \frac{N}{2} > M$$

whenever $N < n \in P$. Consequently $\lim_n n^2/(n + 1) = +\infty$.

Next we take the sequence f so defined that $f_n = -\infty$ whenever n is an odd positive integer and $f_n = n - n^{3/2}$ whenever n is an even positive integer. We see that if n is even, $n \geq 2$ and

$$f_n = n(1 - \sqrt{n}) \leq (1 - \sqrt{2})n < -0.4n$$

If M is an arbitrary finite positive number, we may select $N \in P$ so that $N > 2.5\,M$. Then for any even positive integer $n > N$ we have

$$f_n < -0.04n < -0.4N < -M.$$

For any odd positive integer $n > N$ we have $f_n = -\infty < -M$ also; hence $\lim_n f_n = -\infty$.

We shall now prove some of the limit properties of sequences taking values in Re. The first of these results generalizes Theorem 4.9.

5.13 Theorem. If A and B are sequences with values and limits in Re, $Q \in P$, and $A_n \leq B_n$ whenever $Q \leq n \in P$, then $\lim_n A_n \leq \lim_n B_n$. If, in addition, C is a sequence with values in Re, $\lim_n A_n = \lim_n B_n$, and $A_n \leq C_n \leq B_n$ whenever $Q \leq n \in P$, then $\lim_n C_n$ exists and coincides with the common value of $\lim_n A_n$ and $\lim_n B_n$.

Proof. We let $a = \lim_n A_n$, $b = \lim_n B_n$. If $a = -\infty$ or $b = +\infty$, the statement of the theorem is obviously true. Also, if $a \in R$ and $b \in R$, then as was noted in the proof of Theorem 5.11, there exist positive integers Q' and Q'' such that $A_n \in R$ whenever $Q' < n \in P$, and $B_n \in R$ whenever $Q'' < n \in P$. Therefore $A_n \in R$, $B_n \in R$, and $A_n \leq B_n$ whenever $N < n \in P$, where $N = Q + Q' + Q''$. From this point on we may take over the proof of Theorem 4.9 to infer that $a \leq b$.

The only cases remaining to be considered are those for which $a = +\infty$ or $b = -\infty$. Now given an arbitrary finite number $M > 0$, there exists $Q''' \in P$ such that

$$A_n > M \quad \text{or} \quad B_n < -M,$$

respectively, whenever $Q''' < n \in P$. Thus if $Q + Q''' < n \in P$, we will have

$$B_n \geq A_n > M \quad \text{or} \quad A_n \leq B_n < -M,$$

respectively, leading to $\lim_n B_n = +\infty$ or $\lim_n A_n = -\infty$, respectively. In either case, we see that $\lim_n A_n = \lim_n B_n$.

We now prove the last part of the theorem. If $a = b \in R$, then it follows that there exists a positive integer L such that A and B are finite real numbers whenever $L < n \in P$. The desired conclusion follows in this case by taking over the proof of the corresponding part of Theorem 4.9.

If $a = b = +\infty$ $(-\infty)$, then given an arbitrary finite positive number M, there exists $K \in P$ such that

$$M < A_n, \quad (B_n < -M)$$

whenever $K < n \in P$. Thus

$$M < A_n \leq C_n \leq B_n, \quad (A_n \leq C_n \leq B_n < -M)$$

whenever $K + Q < n \in P$. Thus $\lim_n C_n = +\infty$ $(-\infty)$ as required.

5.14   Lemma.   If $b \in Re$, A is a sequence with values in Re, and A is essentially constantly equal to b (cf. Definition 4.10), then $\lim_n A_n = b$.

Proof.   If $b \in R$, the desired conclusion may be obtained by taking over the proof of Lemma 4.11. If $b = +\infty$ $(-\infty)$, then given any finite positive number M, there exists $N \in P$ such that $A_n = b$ whenever $N \leq n \in P$, and consequently

$$A_n = b > M, \quad (A_n = b < -M)$$

for each such n; therefore $\lim_n A_n = b$.

5.15   Corollary.   If A is a sequence with values in Re, $\lim_n A_n \in Re$, $b \in Re$, $Q \in P$, and $b \leq A_n$ ($A_n \leq b$) whenever $Q \leq n \in P$, then $b \leq \lim_n A_n$ ($\lim_n A_n \leq b$).

Proof.   We define $B = \{b\}$, so that B is a constant sequence, and apply Theorem 5.13.

5.16   Theorem.   If A and B are sequences with values in Re, $a = \lim_n A_n \in R$, $b = \lim_n B_n \in R$, then

(i) $\lim_n |A_n| = |\lim_n A_n|$.

(ii) $\lim_n kA_n = k \lim_n A_n$ if $k \in R$.

(iii) $\lim_n (A_n + B_n) = \lim_n A_n + \lim_n B_n$.

(iv) $\lim_n A_n \cdot B_n = (\lim_n A_n) \cdot (\lim_n B_n)$.

(v) $\lim_n (A_n - B_n) = \lim_n A_n - \lim_n B_n$.

(vi) if $a > 0$ $(a < 0)$ then there exists $N \in P$ such that $A_n > a/2$ $(A_n < a/2)$ whenever $N < n \in P$.

(vii) $\lim_n (A_n/B_n) = (\lim_n A_n)/(\lim_n B_n)$ if $b \neq 0$.

Proof.   The fact that a and b belong to R requires that there exist $Q \in P$ such that $A_n \in R$ and $B_n \in R$ whenever $Q < n \in P$. The desired conclusions follow by taking over the proofs of the various theorems of Chapter 4 between 4.13 and 4.22 inclusive.

5.17   Theorem.   If A and B are sequences with values in Re, $\lim_n A_n \in R$ and $\lim_n B_n = +\infty$ $(-\infty)$, then $\lim_n (A_n + B_n) = \lim_n A_n + \lim_n B_n = +\infty$ $(-\infty)$.

Proof.   We let $a = \lim_n A_n$. Given an arbitrary finite positive number M, there exist positive integers Q and Q' such that

$$B_n > M + |a| + 1, \qquad (B_n < -(M + |a| + 1))$$

whenever $Q < n \in P$ and $|A_n - a| < 1$, hence $|A_n| < |a| + 1$, whenever $Q' < n \in P$. Thus if $Q + Q' < n \in P$, we have

$$A_n + B_n > (M + |a| + 1) - (|a| + 1) = M$$
$$(A_n + B_n < -(M + |a| + 1) + (|a| + 1) = -M);$$

consequently $\lim_n (A_n + B_n) = +\infty$ $(-\infty)$. Reference to § 5.10 completes the proof.

5.18   Theorem.   If A and B are sequences with values in Re, and $\lim_n A_n = \lim_n B_n = +\infty\,(-\infty)$, then $\lim_n (A_n + B_n) = \lim_n A_n + \lim_n B_n = +\infty$ $(-\infty)$.

Proof.   Given an arbitrary finite positive number M, there exist positive integers Q and Q' such that $A_n > M/2$ $(A_n < -M/2)$ whenever $Q < n \in P$, and $B > M/2$ $(B < -M/2)$ whenever $Q' < n \in P$. Thus we see that

$$A_n + B_n > M, \qquad (A_n + B_n < -M)$$

whenever $Q + Q' < n \in P$, and consequently $\lim_n (A_n + B_n) = +\infty$ $(-\infty)$. Reference to § 5.10 completes the proof.

5.19   Lemma.   If A is a sequence with values in Re and $\lim_n A_n = +\infty$ $(-\infty)$, then $\lim_n (-A_n) = -\lim_n A_n$.

Proof.   Given any finite positive number M, there exists $N \in P$ such that

$$A_n > M, \qquad (A_n < -M)$$

whenever $N < n \in P$; hence for each such n,

$$-A_n < -M, \qquad (-A_n > M)$$

whence $\lim_n (-A_n) = -\infty \, (+\infty)$. Reference to § 5.10 completes the proof

5.20   Theorem.   If A and B are sequences with values in Re, and $0 \neq \lim_n A_n \in Re$, $\lim_n B_n = +\infty \, (-\infty)$, then $\lim_n A_n \cdot B_n = (\lim_n A_n) \cdot (\lim_n B_n)$.

Proof.   We let $a = \lim_n A_n$. We consider first the case in which $0 < a \in Re$.

If $a = +\infty$ then there exists $N \in P$ such that $A_n > 1$ whenever $N < n \in P$; and given any finite positive number M there exists $N' \in P$ such that

$$B_n > M, \qquad (B_n < -M)$$

whenever $N' < n \in P$. Thus if $N + N' < n \in P$ we have

$$A_n \cdot B_n > M, \qquad (A_n \cdot B_n < -M)$$

and so $\lim_n A_n \cdot B_n = +\infty \, (-\infty)$. Reference to § 5.10 yields the desired conclusion.

If $a \in R$, then by Theorem 5.16 (vi) there exists $Q \in P$ such that $A_n > a/2$ whenever $Q < n \in P$. Given M as before, there exists $Q' \in P$ such that

$$B_n > 2\frac{M}{a}, \qquad \left(B_n < -2\frac{M}{a}\right)$$

whenever $Q' < n \in P$; thus if $Q + Q' < n \in P$, then

$$A_n \cdot B_n > \frac{2M}{a} \cdot \frac{a}{2} = M \qquad \left(A_n \cdot B_n < -\frac{2M}{a} \cdot \frac{a}{2} = -M\right)$$

regardless of whether A_n and B_n are finite or not, whence $\lim_n A_n \cdot B_n = +\infty \, (-\infty)$. Again reference to § 5.10 yields the desired conclusion.

If $a < 0$, then $-a > 0$, and using the result just derived and Lemma 5.19 we see that $\lim_n (A_n \cdot B_n) = -\lim_n (-(A_n \cdot B_n)) = -\lim_n ((-A_n) \cdot (B_n)) = -(\lim_n (-A_n)) \cdot (\lim_n B_n) = (\lim_n A_n) \cdot (\lim_n B_n)$. This completes the proof.

5.21 **Corollary.** If A is a sequence with values in Re, $\lim_n A_n = +\infty \, (-\infty)$, and $0 \neq k \in Re$, then $\lim_n kA_n = k \lim_n A_n$.

Proof. Let B be the constant sequence $\{k\}$ and apply Theorem 5.20.

5.22 **Theorem.** If A is any sequence with values in Re, $N \in P$, $0 \leq K \in R$, $|A_n| \leq K$ whenever $N \leq n \in P$, and $\lim_n B_n = +\infty \, (-\infty)$ then $\lim_n A_n/B_n = 0$.

Proof. Given an arbitrary finite positive number ϵ, then there exists $Q \in P$ such that

$$B_n > \frac{K+1}{\epsilon}, \qquad \left(B_n < -\frac{K+1}{\epsilon}\right)$$

that is,

$$|B_n| > \frac{K+1}{\epsilon}, \qquad \text{whenever } Q < n \in P.$$

Then if $N + Q < n \in P$ we have $|A_n/B_n| \leq K/|B_n| < \epsilon K/(K+1) < \epsilon$, whence $\lim_n (A_n/B_n) = 0$.

5.23 **Theorem.** If A is a sequence taking values in Re, $N \in P$, $0 < A_n \, (A_n < 0)$ for all n, $N < n \in P$, and $\lim_n 1/A_n = 0$, then $\lim_n A_n = +\infty \, (-\infty)$.

Proof. We suppose given an arbitrary finite positive number M. Then there exists $Q \in P$ such that $|1/A_n| < 1/M$ whenever $Q < n \in P$. Thus $M < A_n \, (A_n < -M)$ whenever $N + Q < n \in P$; consequently $\lim_n A_n = +\infty \, (-\infty)$.

Definition 4.23 and Lemmas 4.24 and 4.25 are applicable and valid for sequences with values in Re, as may be observed easily. Accordingly, we shall accept without further ado the results of these two theorems as applied to sequences in Re. The following is an extension of the results achieved in Theorem 4.26.

5.24 **Theorem.** If A is a non-decreasing (non-increasing) sequence with values in Re, then $\lim_n A_n \in Re$; in fact, $\lim_n A_n = \sup \mathscr{R}(A)$ $(\lim_n A_n = \inf \mathscr{R}(A))$.

Proof. We let $S = \mathscr{R}(A)$, $y = \sup \mathscr{R}(A)$ $(y = \inf \mathscr{R}(A))$.

If $y = -\infty$ $(y = +\infty)$, then evidently $A_n \leq -\infty$ $(A_n \geq +\infty)$ for each $n \in P$; whence $A = \{-\infty\}$ $(A = \{+\infty\})$ is a constant sequence, and $\lim_n A_n = -\infty$ $(+\infty)$ as required.

By Theorem 5.5 if $y = +\infty$ $(y = -\infty)$ and M is any finite positive number, then there exists $x \in S$ such that $M < x \leq y = +\infty$ $(-\infty = y \leq x < -M)$. Also there exists $Q \in P$ such that $x = A_Q$. By virtue of Lemma 4.24, if $Q < n \in P$, then $M < A_Q \leq A_n$ $(A_n \leq A_Q < -M)$, and so $\lim_n A_n = y = +\infty$ $(-\infty)$.

If $y \in R$, then by Theorem 5.5, for any $\epsilon > 0$ there exists $x \in S$ such that $y - \epsilon < x \leq y$ $(y \leq x < y + \epsilon)$. Also there exists $N \in P$ such that $x = A_N$. Due to Lemma 4.24 we have then

$$y - \epsilon < A_N \leq A_n \leq y \quad (y \leq A_n \leq A_N < y + \epsilon)$$

whenever $N < n \in P$, that is, $|A_n - y| < \epsilon$ for each such n, and so $\lim_n A_n = y$ as required.

We next consider an arbitrary sequence A with values in Re. For each $p \in P$, we define

$$\mathscr{A}_p = \{x \mid \text{there exists } n \text{ such that } p \leq n \in P \text{ and } x = A_n\}.$$

We see at once that if $p \in P$, then given $x \in \mathscr{A}_{p+1}$ there exists a positive integer $n \geq p + 1$ for which $x = A_n$; and since necessarily $n \geq p$, it follows that $x \in \mathscr{A}_p$. Thus $\mathscr{A}_{p+1} \subset \mathscr{A}_p$ whenever $p \in P$. Consequently, by Theorem 5.4, the sequences $\{\sup \mathscr{A}_p\}$ and $\{\inf \mathscr{A}_p\}$ have values in Re and are non-increasing and non-decreasing, respectively. These facts, together with Theorem 5.24, justify the following definition.

5.25   Definition.   In the terminology just introduced, we define the *limit superior of the sequence A* as $\lim_n (\sup \mathscr{A}_n)$; we define the *limit inferior of A* as $\lim_n (\inf \mathscr{A}_n)$. These may be written as $\limsup A$ or $\limsup_n A_n$, and $\liminf A$ or $\liminf_n A_n$, respectively.

To help the reader gain an appreciation of the concepts just defined, we consider the following sequence A, so defined that

$$A_n = (-1)^{n/2} \text{ if } n \text{ is an even positive integer};$$

$$A_n = 2 + 2^{-n} \text{ if } n \text{ is an odd positive integer}.$$

Clearly $\sup \mathscr{A}_n = 2 + 2^{-n}$ or $\sup \mathscr{A}_n = 2 + 2^{-(n+1)}$ and $\inf \mathscr{A}_n = -1$ for each $n \in P$, hence $\limsup_n A_n = \lim_n (\sup \mathscr{A}_n) = 2$, and $\liminf_n A_n = \lim_n (\inf \mathscr{A}_n) = -1$.

5.26 Theorem. The sequence A with values in Re has a limit if and only if $\limsup_n A_n = \liminf_n A_n$, and the common value is the limit.

Proof. We assume first that

$$-\infty < \limsup_n A_n = L = \liminf_n A_n < +\infty.$$

We note that for any positive integer p, since $A_{p+1} \in \mathscr{A}_{p+1} \subset \mathscr{A}_p$ then

(1) $$\inf \mathscr{A}_p \leq \inf \mathscr{A}_{p+1} \leq A_{p+1} \leq \sup \mathscr{A}_{p+1} \leq \sup \mathscr{A}_p$$

Also from Definition 5.25, it follows that if ϵ is an arbitrary positive number, then there exist positive integers N and N' such that

(2) $$|\sup \mathscr{A}_n - L| < \epsilon \quad \text{whenever} \quad N < n \in P,$$
$$|\inf \mathscr{A}_n - L| < \epsilon \quad \text{whenever} \quad N' < n \in P.$$

Putting (1) and (2) together we see that if $N + N' < n \in P$, then

$$-\epsilon < \inf \mathscr{A}_n - L \leq A_n - L \leq \sup \mathscr{A}_n - L < \epsilon$$

and so $|A_n - L| < \epsilon$ for each such n. Thus $\lim_n A_n = L$.

We suppose next that

$$-\infty < L = \lim_n A_n < +\infty.$$

Given any positive number ϵ there exists $N \in P$ such that $|A_n - L| < \epsilon/2$ whenever $N < n \in P$, that is,

(3) $$L - \frac{\epsilon}{2} < A_n < L + \frac{\epsilon}{2}$$

for each such n. Thus $L - \epsilon/2$ is a lower bound for $\mathscr{A}_{N+1}$ and $L + \epsilon/2$ is an upper bound for $\mathscr{A}_{N+1}$. From (1) and Lemma 4.24 we infer that

$$L - \frac{\epsilon}{2} \leq \inf \mathscr{A}_{N+1} \leq \inf \mathscr{A}_n \leq \sup \mathscr{A}_n \leq \sup \mathscr{A}_{N+1} \leq L + \frac{\epsilon}{2}$$

whenever $N < n \in P$. Thus $|\inf \mathscr{A}_n - L| < \epsilon/2 < \epsilon$ and $|\sup \mathscr{A}_n - L| < \epsilon/2 < \epsilon$ for each such n, and consequently $\liminf_n A_n = \lim_n (\inf \mathscr{A}_n) = L = \lim_n (\sup \mathscr{A}_n) = \limsup_n A_n$.

Next we assume that $\liminf_n A_n = \limsup_n A_n = +\infty$ $(-\infty)$. Given any finite positive number M, there exists $N \in P$ such that

$$A_n \geq \inf \mathscr{A}_n > M \qquad (A_n \leq \sup \mathscr{A}_n < -M)$$

whenever $N < n \in P$; consequently $\lim_n A_n = +\infty$ $(-\infty)$ as required.

Finally we assume that $\lim_n A_n = +\infty\,(-\infty)$. Then if M is an arbitrary finite positive number, there exists $N \in P$ such that

$$A_n > M + 1, \qquad (A_n < -M - 1)$$

whenever $N < n \in P$. Thus $M + 1$ is a lower bound $(-(M + 1)$ is an upper bound) for $\mathscr{A}_{N+1}$. Due to Lemma 4.24, if $N < n \in P$, then

$$\sup \mathscr{A}_n \geq \inf \mathscr{A}_n \geq \inf \mathscr{A}_{N+1} \geq M + 1 > M$$
$$(\inf_n \mathscr{A}_n \leq \sup \mathscr{A}_n \leq \sup \mathscr{A}_{N+1} \leq -M - 1 < -M)$$

and consequently

$$\operatorname{limsup}_n A_n = \lim_n (\sup \mathscr{A}_n) = \lim_n (\inf \mathscr{A}_n) = \operatorname{liminf}_n A_n = +\infty\,(-\infty).$$

The proof is now complete.

5.27 Theorem. If A is a sequence with values in Re, and B is such a subsequence of A that $b = \lim_n B_n$, then $\operatorname{liminf}_n A_n \leq b \leq \operatorname{limsup}_n A_n$.

Proof. Due to Definition 4.27 there exists a strictly increasing positive integer-valued sequence f such that $B = A \circ f$. Due to Lemma 4.28, $f(n) \geq n$ for each $n \in P$. Thus, using Lemma 4.24, we have

$$(1) \qquad \inf \mathscr{A}_n \leq \inf \mathscr{A}_{f(n)} \leq B_n = A\big(f(n)\big) \leq \sup \mathscr{A}_{f(n)} \leq \sup \mathscr{A}_n$$

and so, by Theorem 5.13,

$$\operatorname{liminf}_n A_n = \lim_n (\inf \mathscr{A}_n) \leq \lim_n B_n = b \leq \lim_n (\sup \mathscr{A}_n) = \operatorname{limsup}_n A_n.$$

5.28 Corollary. If A is any sequence taking values in Re, then A has a limit in Re if and only if every subsequence has a common limit, which is in fact the limit of A.

Proof. Due to Theorems 5.26, 5.13, and (1) in the proof of Theorem 5.27, it follows that if A has a limit, then each subsequence of A has the same limit. Now since A is itself a subsequence of A (cf. proof of Corollary 4.30), then if all subsequences of A have a common limit, A itself has that limit.

We note from Theorem 5.27 that the numbers $\operatorname{limsup}_n A_n$ and $\operatorname{liminf}_n A_n$ are the largest and smallest limit values respectively that may be taken by any subsequence of A. This gives some justification for the names "limit superior" and "limit inferior." As we will show in Theorems 6.6 and 6.7, there exist subsequences of A that take the values $\operatorname{limsup}_n A_n$ and $\operatorname{liminf}_n A_n$ respectively, as limits, which strengthens the argument for using these names.

We shall agree herewith to extend Definition 4.32 to apply to sequences with values in *Re* and to name any such sequence complying with its terms a *Cauchy sequence*. We are able to obtain the following improvement on Corollary 4.36 under this slightly more general situation.

5.29 Theorem.   The sequence *A* with values in *Re* converges to a (finite) limit if and only if it is a Cauchy sequence.

Proof.   If *A* is a Cauchy sequence, an examination of the proof of Theorem 4.34 reveals easily that there exists $M \in P$ such that $(A \mid (P - \hat{M}))$ takes only (finite) real values. We define the sequence *B* so that

$$B_n = 0 \text{ if } n \in \hat{M}, \qquad B_n = A_n \text{ if } n \in (P - \hat{M}),$$

check that *B* is a (finite) real-valued Cauchy sequence, infer that *B* converges to a finite limit by Theorem 4.34, and invoke Lemma 5.12 to conclude that *A* converges to this same limit.

The proof of Theorem 4.35 may be followed verbatim to establish the converse.

There is no counterpart to the Cauchy criterion for infinite limits. This is one reason for reserving the name *convergent* for sequences with finite limits.

Exercises

§§ 5.1 to 5.12

Prove the following statements from the definition of infinite limits.

1. $\lim_n n^2/(3n + 20) = +\infty$.

2. $\lim_n (n^3 - n)/(n^2 + 1) = +\infty$.

3. $\lim_n (n^3 - 2n)/(5n^2 + 100) = +\infty$.

4. $\lim_n (10 - n^2)/(n - 50) = -\infty$.

5. $\lim_n x^n = +\infty$ if $1 < x \in R$. (*Hint:* let $x = 1 + h$, $h > 0$, and recall that $x^n > 1 + nh$ whenever $n \in P$).

6. $\lim_n \sqrt{n} = +\infty$. (*Hint:* See Problem 14 in the Exercises at the end of Chapter 4.)

7. $\lim_n \sqrt[k]{n} = +\infty$ whenever $k \in P$. (*Hint:* See Problem 15 in the Exercises at the end of Chapter 4.)

8. $\lim_n n! = +\infty$.

9. $\lim_n \dfrac{n^n}{n!} = +\infty$.

10. $\lim_n \dfrac{n!}{x^n} = +\infty$ if $0 < x \in R$.

11. $\lim\limits_{n} \dfrac{n^5 - 1}{10n^3 + 1} = +\infty.$

12. $\lim\limits_{n} \dfrac{-n^3 + 12n^2 + n + 100}{75n^2 - 50} = -\infty.$

13. $\lim\limits_{n} \dfrac{-n^2 + 1000}{n + 1} = -\infty.$

14. If $k \in P,\, p \in P,\, k < p,\, b_0 \neq 0,\, a_0/b_0 > 0$ $(a_0/b_0 < 0)$, then

$$\lim\limits_{n} \frac{a_0 n^p + a_1 n^{p-1} + \ldots + a_p}{b_0 n^k + b_1 n^{k-1} + \ldots + b_k} = +\infty(-\infty).$$

15. Show that if $x \in Re,\, y \in Re$, and $x < y$, then there exists $z \in Re$ such that $x < z < y$. (*Hint:* Since the statement holds if x and y are finite real numbers, we need only show this if $x = -\infty$, or $y = +\infty$, or both relations hold.)

16. Show that if $S \subset Re$ and $T = \{x \mid -x \in S\}$, then $\sup T = -\inf S$ and $\inf T = -\sup S$.

§§ 5.13 to 5.29

17. Given $A = \{n^2\},\, B = \{n\},\, C = \{n^3\}$, show that $\lim\limits_{n} A_n = \lim\limits_{n} B_n = \lim\limits_{n} C_n = +\infty$. Show that $\lim\limits_{n}(A_n - B_n) = +\infty,\, \lim\limits_{n}(A_n - C_n) = -\infty,\, \lim\limits_{n} A_n/B_n = +\infty,\, \lim\limits_{n} A_n/C_n = 0.$

From these examples, do you think it would be possible to give sensible definitions to $+\infty - (+\infty)$ or $+\infty/+\infty$?

18. Given the sequence f such that

$$f(n) = (-1)^n \cdot n,$$

prove whether or not f has a limit. (*Hint:* Recall Corollary 5.28.)

19. Given $f(n) = \dfrac{n^3 - 1000n^2}{(-1)^n}$, prove whether or not f has a limit.

20. Prove whether or not $\lim\limits_{n} \dfrac{n!}{n^4 + 1}$ exists.

21. Prove whether or not $\lim\limits_{n} \dfrac{x^n}{n^{10}}$ exists if $|x| > 1$. (*Hint:* Consider two cases, $x > 1,\, x < -1$.)

22. Evaluate $\lim\limits_{n} (5n^6 - 7(-1)^n n^4)$ if this limit exists. Otherwise prove that it does not exist.

23. Evaluate $\lim\limits_{n} (2^n - (-2)^n)$ if this limit exists. Otherwise prove that it does not exist.

24. Prove whether or not $\lim\limits_{n} (n! - 2^n)$ exists.

25. Prove whether or not $\lim\limits_{n} (3^n - (-2)^n)$ exists.

26. Prove that $\lim\limits_{n} \dfrac{2^n}{n} = +\infty.$

27. Let f be a non-negative sequence, $N \in P$, and $r > 1$ such that $\dfrac{f(n + 1)}{f(n)} > r$ for each n such that $N \leq n \in P$. Prove that $\lim\limits_{n} f(n) = +\infty$. (*Hint:* Show first that $f(m + N) > r^m f(N)$ for each $m \in P$ by induction.)

28. Given the sequence A so defined that

$$A(n) = 1 + (\tfrac{1}{2})^n \text{ if } n \text{ is an even positive integer}$$

$$A(n) = -\frac{1}{n} \text{ if } n \text{ is an odd positive integer,}$$

determine $\sup \mathcal{A}_n$ and $\inf \mathcal{A}_n$ for each $n \in P$. Then determine $\limsup_n A_n$ and $\liminf_n A_n$.

29. Given the sequence A so defined that

$$A(n) = n^2 - n \text{ if } n \text{ is an even positive integer}$$

$$A(n) = 1 - \frac{1}{n} \text{ if } n \text{ is an odd positive integer,}$$

find $\sup \mathcal{A}_n$ and $\inf \mathcal{A}_n$ for each $n \in P$ and then determine $\limsup_n A_n$ and $\liminf_n A_n$.

30. Let A be so defined that

$$A(n) = -n^2 + 1 \text{ whenever } n \text{ is divisible by } 3$$

$$A(n) = -2 + \frac{1}{n} \text{ whenever } n = 3k - 1 \text{ for some } k \in P$$

$$A(n) = n + 2 \text{ whenever } n = 3k - 2 \text{ for some } k \in P.$$

Find $\sup \mathcal{A}_n$ and $\inf \mathcal{A}_n$ for each $n \in P$ and then determine $\limsup_n A_n$, $\liminf_n A_n$.

31. Let A be so defined that

$$A(n) = \frac{2}{n^2} + 1 \text{ whenever } n \text{ is divisible by } 4$$

$$A(n) = (-1)^n n \text{ whenever } n = 4k - 1 \text{ for some } k \in P$$

$$A(n) = +\frac{1}{n} \text{ whenever } n = 4k - 2 \text{ for some } k \in P$$

$$A(n) = (-1)^n \text{ whenever } n = 4k - 3 \text{ for some } k \in P.$$

Find $\liminf_n A_n$ and $\limsup_n A_n$

32. Let A be so defined that $A(1) = 1$, $A(2) = \tfrac{1}{2}$, $A(3) = \tfrac{2}{2}$, $A(4) = \tfrac{1}{3}$, $A(5) = \tfrac{2}{3}$, $A(6) = \tfrac{3}{3}$, etc. Determine $\liminf_n A_n$ and $\limsup_n A_n$.

33. Show that if $S \subset Re$ and a is a cluster point of S, then a is a cluster point of each set T such that $S \subset T \subset Re$.

34. Show that the result of Theorem 5.16 (i) holds in case $a = +\infty$ or $a = -\infty$.

35. Show that if $\varnothing \neq A \subset Re$ and A has a smallest member a, then $a = \inf A$; if A has a largest member b, then $b = \sup A$.

36. Show that every infinite subset of Re has a cluster point. (*Hint:* Consider two separate cases, one involving subsets that have a finite upper and lower bound, the other those that do not. Note that this theorem is not true for the real number system itself.)

6

Definition by Induction

Whenever we define a set, we must give a clear-cut statement enabling us to tell just which objects in the mathematical universe are its members. As we saw in Chapter 1, functions are merely special kinds of sets, so that this principle is applicable to them. Alternatively and equivalently, we may define a function f by expressing explicitly the value $f(x)$ associated with an arbitrary member x of its domain, which must itself be defined unambiguously.

Sometimes, however, we see a function f defined on the positive integers in the following way: $f(1)$ is prescribed; then $f(2)$ is defined in some clearcut fashion that depends on the value $f(1)$; then $f(3)$ is defined unambiguously in a manner that depends on the value $f(2)$ and (at least indirectly) on the value $f(1)$. In general, a rule is given that determines $f(n + 1)$ in terms of the values $f(1), f(2), \ldots, f(n)$, for any positive integer n. Without further ado, it is frequently taken for granted that this process defines uniquely a function f whose domain is the set P of all positive integers. The method is known as *definition by induction*. We propose to investigate its legitimacy.

At first glance it may seem that there is nothing to investigate, for it is intuitively clear that if we are given any specific positive integer, then such a procedure may be applied an appropriate number of times to obtain eventually the desired function value. For instance, suppose we wish to define a function f on P so that $f(1) = 2$ and $f(n + 1) = 2 \cdot f(n)$ for each $n \in P$. We see at once that for this particular function f, we must have $f(2) = 2 \cdot f(1) = 2 \cdot 2 = 4$, $f(3) = 2 \cdot f(2) = 2 \cdot 4 = 8$, etc. Given a specific positive integer k, say $k = 5$, or $k = 27$, or $k = 2492$, we can see intuitively, and even actually if we wish to take the time, that the value $f(k)$ can be and is determined uniquely by repeating a sufficient number of times the operation supposedly defining the function in question. However, this still does not prove that there exists any function f whose domain is the infinite set P satisfying the relation $f(1) = 2$, $f(n + 1) = 2 \cdot f(n)$ for each $n \in P$, to say nothing of its uniqueness. The situation here is much like that described and discussed in § 2.4, with regard to proof by induction.

The necessity for justification of definition by induction is apparent when we subject the process to the tests mentioned earlier. We see that it does not tell us exactly which ordered pairs belong to the function. Furthermore, for an arbitrary positive integer $n > 1$, the value of $f(n)$ is not given explicitly, but instead is expressed somehow in terms of $f(1), f(2), \ldots, f(n-1)$. In other words, f appears to be defined in terms of itself, which is a violation of a basic principle.

This seeming difficulty can be circumvented simply by looking at the problem from a somewhat different point of view. Instead of saying: let f be so defined that $f(1)$ is a certain prescribed thing, and for each $n \in P, f(n+1)$ is defined in some prescribed fashion in terms of the values $f(1), f(2), \ldots, f(n)$, we may pose the matter in the form of a question. We ask: does there exist a function f defined on P such that f satisfies the given conditions? If some such function f exists, is it unique? From this standpoint, the problem becomes one of ascertaining by ordinary logic the satisfiability of a certain functional relationship, and later that the solution is unique. For a large class of such functional relationships, including in particular all those encountered in elementary mathematics, it turns out that there exist unique solutions. In each of these cases the process of definition by induction is justified.

As it happens, though, the proof of this fact does not generally give us either an explicit criterion for determining which ordered pairs belong to the desired function f or the explicit value $f(n)$ associated with an arbitrary $n \in P$. This may seem to leave us just where we started, but such is not the case. The fact that we manage to prove the existence of the desired function by acceptable logic is itself a guarantee of the existence of a criterion for membership of ordered pairs in the function; the fact that we do not always find an explicit form of this criterion is not a defect of the logic of this proof. However, when we are unable to find an expression for f or for its values, then we have only the functional relationship given at the outset as our source of information about f, and any additional information that we may obtain about f has to be derived from this relationship. As we shall see in applications, it is usually necessary to use proof by induction to gain such additional information.

We turn now to the problem of justifying definition by induction. We shall denote by $\mathscr{P}$ the class of all functions f for which there exists $n \in P$ such that $\mathscr{D}(f) = \hat{n}$. This is truly a gigantic class. It should be borne in mind that the members of $\mathscr{P}$ do not have to be numerical valued.

Now suppose that we wish to define a function f on P, such that $f(1)$ is some prescribed mathematical object α, and such that $f(n+1)$ is defined by an unambiguous rule in terms of $f(1), f(2), \ldots, f(n)$, for each $n \in P$. We seek to discover what conditions, as a minimum, we might expect to have to impose on such a given rule in order to hope to obtain an answer to our problem. Now the values $f(1), f(2), \ldots, f(n)$, are determined by and in turn determine the function $(f \mid \hat{n})$. Hence our rule effectively associates with the function $(f \mid \hat{n})$ the value $f(n + 1)$. Thus it appears that our rule behaves like a function, and so is a function, whose value at $(f \mid \hat{n})$ is precisely $f(n + 1)$. Let us denote this hypothetical function by F; our first condition on F is that for each $n \in P$, we must have $F((f \mid \hat{n})) = f(n + 1)$.[1] It is also apparent that the domain of F must be a class of functions that includes all functions of the form $(f \mid \hat{n})$, where $n \in P$, and f is the solution (if it exists) of our problem. Since we cannot know in advance what the function f is going to be, in case it does exist, then we have to endow F with a domain sufficiently large so that it is bound to include $(f \mid \hat{n})$ for each $n \in P$, regardless of what f happens to be. To ensure this, we need to insist that $\mathscr{D}(F) = \mathscr{P}$. Thus we see that, as a minimum requirement, the rule that determines the value of $f(n + 1)$ for each $n \in P$ must itself be a function F with $\mathscr{D}(F) = \mathscr{P}$. As we shall see, this is the only restriction we have to impose on F. We are now prepared to prove the basic theorem on definition by induction.

6.1   Theorem. If F is any function with $\mathscr{D}(F) = \mathscr{P}$, and α is any prescribed mathematical object, then there exists a function f with $\mathscr{D}(f) = P$, $f(1) = \alpha$, and $f(n + 1) = F(f \mid \hat{n})$ for each $n \in P$. Furthermore, f is unique.

Proof. We begin by solving a simpler problem, namely, for an arbitrary positive integer n, we shall show that there exists a function g with $\mathscr{D}(g) = \hat{n}$ such that

   (i) $g(1) = \alpha$;
   (ii) $g(i + 1) = F(g \mid \hat{i})$ whenever $i \in \hat{m}$, where $m = n - 1$ if $n > 1$, using proof by induction.[2] We let Q be the set of all positive integers n for which there exists g satisfying (i) and (ii). We let $h_0 = \langle (1, \alpha) \rangle$; clearly, h_0 is a function satisfying (i), and (ii) is satisfied vacuously by h_0. Thus $1 \in Q$.

We now assume that $n \in Q$, and show that $n + 1 \in Q$ follows from this assumption. We assume g satisfies (i) and (ii) and wish to prove the existence of a function h such that $\mathscr{D}(h) = \hat{p}$, where $p = n + 1$, and

(1) $$h(1) = \alpha;$$

(2) $$h(i + 1) = F(h \mid \hat{i}) \quad \text{whenever} \quad i \in \hat{n}.$$

The only difference between (ii) and (2) is that (2) has to be satisfied for one additional value of i, namely $i = n$. This suggests that one might

---

[1] Henceforth for the sake of brevity we will write $F(f \mid \hat{n})$ instead of $F((f \mid \hat{n}))$, etc.
[2] This is legitimate; *proof* by induction does not presuppose *definition* by induction.

simply try to extend g. Let us then so define a function h on $\hat{p}$ that $h(i) = g(i)$ for each $i \in \hat{n}$. To complete the job we need only look for a way of defining h at $n + 1 = p$ so as to satisfy

$$(3) \qquad\qquad h(n + 1) = F(h \mid \hat{n}).$$

However, since $g = (h \mid \hat{n})$ due to our definition thus far, we can satisfy (3) by simply defining $h(n + 1) = F(g)$, and consequently this is what we do. Then $\mathscr{D}(h) = \hat{p}$, and h satisfies the conditions (1) and (2) for each $i \in \hat{n}$. Consequently $n + 1 \in Q$, and $Q = P$.

We now let $\mathscr{F}$ denote the family of those functions g for which there exists $n \in P$ such that $\mathscr{D}(g) = \hat{n}$ and conditions (i) and (ii) are valid. We will show $\mathscr{F}$ is a nest (recall Definition 1.49). Accordingly, we consider any two members g and g' of $\mathscr{F}$. There exist positive integers n and n' such that $\mathscr{D}(g) = \hat{n}$, $\mathscr{D}(g') = \hat{n}'$, and the conditions (i) and (ii) are valid for g and g'. We must show that $g \subset g'$ or $g' \subset g$. Now either $n \leq n'$ or $n' \leq n$. We will assume that $n \leq n'$ and show that in this case $g \subset g'$; if $n' \leq n$, then in similar fashion $g' \subset g$.

We let

$$S = \{i \mid i \in \hat{n} \quad \text{and} \quad g(i) \neq g'(i)\}.$$

It follows from Theorem 1.39 that $g \not\subset g'$ only if $S \neq \varnothing$. Let us assume, if possible, that $S \neq \varnothing$; then S has a smallest member j. Since $g(1) = \alpha = g'(1)$ by (i), then $j > 1$ and so $k = j - 1 \in P$. From our definition of j, we have $g(i) = g'(i)$ for each $i \in \hat{k}$, that is, $(g \mid \hat{k}) = (g' \mid \hat{k})$. From (ii) we see that

$$g(j) = g(k + 1) = F(g \mid \hat{k}) = F(g' \mid \hat{k}) = g'(k + 1) = g'(j).$$

This contradicts the fact that $j \in S$ and establishes that $S = \varnothing$ and $g \subset g'$. Thus $\mathscr{F}$ is a nest.

We now let $f = \bigcup \mathscr{F}$. By virtue of Theorem 1.50, f is a function. We will show that f is the solution to our problem. We see first that if n is an arbitrary positive integer, then from the first part of our proof, there exists a function $g \in \mathscr{F}$ with $\mathscr{D}(g) = \hat{n}$. Now $g \subset f$ by Theorem 1.21; therefore $\hat{n} = \mathscr{D}(g) \subset \mathscr{D}(f)$; hence $n \in \mathscr{D}(f)$. Thus $P \subset \mathscr{D}(f)$. Also, if $x \in \mathscr{D}(f)$, then there exists y such that $(x, y) \in f$, and there exists $g \in \mathscr{F}$ such that $(x, y) \in g$, whence $x \in \mathscr{D}(g) \subset P$. Thus $\mathscr{D}(f) \subset P$, and so we conclude that $\mathscr{D}(f) = P$.

As we saw earlier, $\langle (1, \alpha) \rangle \in \mathscr{F}$; hence $(1, \alpha) \in \bigcup \mathscr{F} = f$. Consequently $f(1) = \alpha$. Next we consider an arbitrary positive integer n. There exists a function $g \in \mathscr{F}$ with $\mathscr{D}(g) = \hat{p}$, where $p = n + 1$. Since $g \subset f$, then $(g \mid \hat{n}) = (f \mid \hat{n})$ and $g(n + 1) = f(n + 1)$. Using (ii) we conclude that

$$f(n + 1) = g(n + 1) = F(g \mid \hat{n}) = F(f \mid \hat{n}).$$

To complete our theorem, we have only to show that if f and f' are functions on P satisfying the requirements of the theorem, then $f = f'$. We consider

$$T = \{i \mid i \in P \text{ and } f(i) \neq f'(i)\}.$$

If $f \neq f'$, then $T \neq \varnothing$ and so T must have a smallest member j. Since $f(1) = \alpha = f'(1)$ then $k = j - 1 \in P$ and $(f \mid \hat{k}) = (f' \mid \hat{k})$ due to our definition of j. By our assumptions on f and f' we conclude that

$$f(j) = f(k + 1) = F(f \mid \hat{k}) = F(f' \mid \hat{k}) = f'(k + 1) = f'(j),$$

contradicting the fact that $j \in T$. Thus we infer that $T = \varnothing$ and $f = f'$.

The theorem just proved shows that each function F with domain $\mathscr{P}$ may be regarded as generating a unique function f, with prescribed initial value, that may be thought of as defined inductively by F. To legitimize the definitions of some of the common functions, such as positive integral powers of a given real number and factorials, it is necessary to find appropriate functions F that generate the functions in question.

As an example of this, suppose we wish to legitimize the function f for which $f(1) = 2$, $f(n + 1) = 2 \cdot f(n)$ for each $n \in P$. This is the well-known function f such that $f(n) = 2^n$. We have to define a function F on $\mathscr{P}$ to generate f. Since the class $\mathscr{P}$ contains all sorts of non-numerical-valued functions which are of no interest to us here, we usually find it convenient to define F in some simple fashion on these members of $\mathscr{P}$. Thus we dispose of this part of the definition of F by setting $F(g) = 0$ whenever $g \in \mathscr{P}$ and g is not numerical valued.

It is the values taken by F on the remainder of $\mathscr{P}$ that command our attention. In accordance with Theorem 6.1 and the special requirements we want our function f to satisfy, we must have

$$2 \cdot f(n) = f(n + 1) = F(f \mid \hat{n})$$

for each $n \in P$. This suggests that we so define F that for each numerical-valued function g in $\mathscr{P}$,

$$F(g) = 2 \cdot g(\sup \mathscr{D}(g)).$$

This completes the definition of F. Now by virtue of Theorem 6.1 we know that F generates a function f on P such that $f(1) = 2$, and also $f(n + 1) = F(f \mid \hat{n})$ for each $n \in P$. All we have to check is that

$$f(n + 1) = 2f(n)$$

for each $n \in P$. But $f(n + 1) = F(f \mid \hat{n}) = 2(f \mid \hat{n})(x)$, where $x = \sup \mathscr{D}(f \mid \hat{n})$, due to the definition of F. Also, $\sup \mathscr{D}(f \mid \hat{n})$ is clearly n itself, and $(f \mid \hat{n})(n) = f(n)$. We see now that the condition we set out to establish is satisfied.

The theorem on definition by induction depends on the well-ordering principle for its proof. It can be used to define functions inductively on any well-ordered set by making slight modifications in the proof.

For our first application of Theorem 6.1, we consider an infinite subset A of P. It seems clear intuitively that it should be possible to find a sequence f whose values $f(1), f(2), f(3)$, etc., give in turn the members of A in increasing order. Intuitively this can be accomplished as follows. We let $f(1)$ be the smallest member of A. Then $A - \langle f(1) \rangle$ is an infinite subset of P, which in turn has a smallest member that we take as $f(2)$; clearly $f(1) < f(2)$. Then $A - (\langle f(1) \rangle \cup \langle f(2) \rangle)$ is an infinite subset of P that has a smallest member that we take as $f(3)$, and so on. To show the existence of the sequence we seek, it is necessary to formalize this procedure. It is clear that definition by induction is involved, since for each $n \in P, f(n + 1)$ is determined by the values $f(1), f(2), \ldots, f(n)$.

6.2 Definition. We define the function L so that for each set $B \subset P$,

$$L(B) = \text{the smallest member of } B \text{ if } B \neq \varnothing$$
$$L(\varnothing) = 0.$$

Because of the well-ordering principle for the positive integers, L is well-defined, and clearly $L(B) \in B \subset P$ whenever $\varnothing \neq B \subset P$. We could choose $L(\varnothing)$ in any way whatever, but it is convenient to define it so that its value is distinct from any other value taken by L.

6.3 Theorem. If A is an infinite subset of P, then there exists a strictly increasing sequence f with $\mathscr{D}(f) = P, f(1) = L(A)$, and $\mathscr{R}(f) = A$.
Proof. We define F on $\mathscr{P}$ so that

$$F(g) = L(A - \mathscr{R}(g)) \quad \text{whenever} \quad g \in \mathscr{P}.$$

Since $L(A - \mathscr{R}(g))$ is either 0 or a positive integer, depending on whether $A - \mathscr{R}(g) = \varnothing$ or not, then F is clearly well-defined on $\mathscr{P}$. From Theorem 6.1 we infer the existence of a unique function f with $\mathscr{D}(f) = P$ such that

(1) $f(1) = L(A); \quad f(n + 1) = F(f \,|\, \hat{n}) = L(A - \mathscr{R}(f \,|\, \hat{n}))$

for each $n \in P$.

Since $\mathscr{D}(f \,|\, \hat{n}) = \hat{n}$, then $\mathscr{R}(f \,|\, \hat{n})$ is a finite set for each $n \in P$ by Definition 3.2; thus $A - \mathscr{R}(f \,|\, \hat{n})$ is an infinite and so non-empty set for each such n. Thus for each n, $1 < n \in P$, it follows that $k = n - 1 \in P$ and

$$f(1) = L(A) \in A$$

(2) $f(k + 1) = f(n) = L(A - \mathscr{R}(f \,|\, \hat{k})) \in (A - \mathscr{R}(f \,|\, \hat{k})) \subset A \subset P$
     $f(n + 1) = L(A - \mathscr{R}(f \,|\, \hat{n})) \in (A - \mathscr{R}(f \,|\, \hat{n})) \subset A.$

Since $(f \mid \hat{k}) \subset (f \mid \hat{n})$ then $\mathscr{R}(f \mid \hat{k}) \subset \mathscr{R}(f \mid \hat{n})$ and so

(3) $$(A - \mathscr{R}(f \mid \hat{n})) \subset (A - \mathscr{R}(f \mid \hat{k})).$$

From the third relation of (2) and (3) we infer that

$$f(n + 1) \in (A - \mathscr{R}(f \mid \hat{k})),$$

whereas $f(k + 1) = f(n)$ is the smallest member of $A - \mathscr{R}(f \mid \hat{k})$. Therefore $f(n) \leq f(n + 1)$. Now $f(n) \in \mathscr{R}(f \mid \hat{n})$, so $f(n) \notin (A - \mathscr{R}(f \mid \hat{n}))$, whereas $f(n + 1) \in (A - \mathscr{R}(f \mid \hat{n}))$ by (2). Thus $f(n) \neq f(n + 1)$. This tells us that $f(n) < f(n + 1)$ whenever $1 < n \in P$. Also, $f(2) \in (A - \mathscr{R}(f \mid 1)) = (A - \langle f(1) \rangle) \subset A$ by (2). From this we see that $f(2) \neq f(1)$; since $f(1)$ is the smallest member of A, then $f(1) < f(2)$. Thus the relation $f(n) < f(n + 1)$ holds for each $n \in P$, and f is strictly increasing.

We have only to show that $\mathscr{R}(f) = A$. By (2), we know that $\mathscr{R}(f) \subset A$. If it were possible that $A \neq \mathscr{R}(f)$, then $(A - \mathscr{R}(f)) \neq \varnothing$ and there must exist $m \in P$ such that $m \in (A - \mathscr{R}(f)) \subset (A - \mathscr{R}(f \mid \hat{m}))$, whence $m \in (A - \mathscr{R}(f \mid \hat{m}))$. Now by (2), $f(m + 1)$ is the smallest member of $A - \mathscr{R}(f \mid \hat{m})$, whence $f(m + 1) \leq m < m + 1$. This contradicts Lemma 4.28. Thus we conclude that $A = \mathscr{R}(f)$ and the proof is complete.

6.4   Corollary.   Every infinite subset A of P is equivalent to P.
Proof.   The mapping function f of Theorem 6.3 is clearly a one-to-one mapping of A onto P, as required by Definition 3.26.

6.5   Corollary.   If A is a non-empty finite subset of P, then there exists $N \in P$ and a strictly increasing function h with $\mathscr{D}(h) = \hat{N}$, $\mathscr{R}(h) = A$.
Proof.   Since A is finite it has a largest member $Q \in P$ by Theorem 3.20. We let $B = A \cup (P - \hat{Q})$. Now $P - \hat{Q}$ is unbounded and so is infinite, thus so is B. Also, the members of A do not exceed Q, whereas all members of $P - \hat{Q}$ do exceed Q. Consequently A and $P - \hat{Q}$ are disjoint. By Theorem 6.3 there exists a strictly increasing sequence f with $\mathscr{D}(f) = P$, $\mathscr{R}(f) = B$. There exists $N \in P$ such that $f(N) = Q$; for any $i \in P$ and $k \in P$, if $i \leq N < k$, then $f(i) \leq f(N) = Q < f(k) \in (P - \hat{Q})$, due to the strictly increasing nature of f. But then $f(i) \in A$ for each such i. Also, to each member x of $A \subset B$ there exists a positive integer j such that $f(j) = x$, and we must have $j \leq N$ by the observation just made. We conclude that the function $h = (f \mid \hat{N})$ is strictly increasing on $\hat{N}$ and $\mathscr{R}(h) = A$, as required.

6.6   Theorem.   If A is a sequence with values in Re, and if $-\infty < \limsup_{n} A_n < +\infty$ $(-\infty < \liminf_{n} A_n < +\infty)$, then there exists a subsequence of A converging to $\limsup_{n} A_n$ $(\liminf_{n} A_n)$.

Proof. We let $M = \limsup\limits_{n} A_n$ ($m = \liminf\limits_{n} A_n$). We so define the sequence T of subsets of P that for each $k \in P$,

$$T_k = \{n \mid k \leq n \in P \quad \text{and} \quad |M - A_n| < 1/k\}$$

$$(T_k = \{n \mid k \leq n \in P \quad \text{and} \quad |m - A_n| < 1/k\}).$$

We suppose $k \in P$, $k' \in P$, $k \leq k'$. If n is an arbitrary member of $T_{k'}$, then $k' \leq n \in P$, and $|M - A_n| < 1/k'$ ($|m - A_n| < 1/k'$). Clearly, then, $k \leq n \in P$ and $|M - A_n| < 1/k' \leq 1/k(|m - A_n| < 1/k' \leq 1/k)$, whence $n \in T_k$. Thus $T_{k'} \subset T_k$.

We assert that for each $k \in P$, T_k is an infinite set. To show this, suppose we are given an arbitrary positive integer L. In the notation associated with Definition 5.25, we see that there exists $N \in P$ such that

(1) $$|M - \sup \mathscr{A}_n| < 1/(2k), \qquad (|m - \inf \mathscr{A}_n| < 1/(2k))$$

whenever $N < n \in P$. We take $q = L + N + k$, and using Theorem 5.5, we note that there exists $x \in \mathscr{A}_q$ such that

(2) $$|x - \sup \mathscr{A}_q| < 1/(2k), \qquad (|x - \inf \mathscr{A}_q| < 1/(2k))$$

Due to the definition of $\mathscr{A}_q$, there exists $p \in P$ such that $p \geq q$ and $x = A_p$. Thus, combining (1) and (2), we obtain $|A_p - M| < 1/k$ ($|A_p - m| < 1/k$); and since $p \geq q > L, p \geq k$, we see that $p \in T_k$ and hence T_k is unbounded above due to the arbitrary nature of L. Thus by Corollary 3.17, T_k is infinite.

Applying Theorem 6.3, we observe that there exists a sequence G of functions such that for each $k \in P$, G_k is a strictly increasing function with $\mathscr{D}(G_k) = P$, $\mathscr{R}(G_k) = T_k \subset P$; in particular, $G_k(1)$ is the smallest member of T_k. We wish to show that if k is an arbitrary member of P, then

$$G_k(i) \leq G_{k+1}(i)$$

for each $i \in P$. We assume the contrary; then for some $k \in P$, the set

$$S = \{i \mid i \in P \text{ and } G_{k+1}(i) < G_k(i)\}$$

is non-empty and so has a smallest member j. Now $G_{k+1}(1) \in \mathscr{R}(G_{k+1}) = T_{k+1} \subset T_k = \mathscr{R}(G_k)$; and since $G_k(1)$ is the smallest member of T_k, then $G_k(1) \leq G_{k+1}(1)$, whence $1 \notin S$, $j > 1$, and $j - 1 \in (P - S)$. Therefore

(3) $$G_k(j - 1) \leq G_{k+1}(j - 1), \qquad G_{k+1}(j) < G_k(j).$$

Since G_{k+1} increases strictly, then $G_{k+1}(j - 1) < G_{k+1}(j)$, and from (3) we obtain

(4) $$G_k(j - 1) < G_{k+1}(j) < G_k(j).$$

However, $G_{k+1}(j) \in \mathscr{R}(G_{k+1}) = T_{k+1} \subset T_k = \mathscr{R}(G_k)$, whence there exists $m \in P$ such that $G_{k+1}(j) = G_k(m)$. From (4) we thus obtain $G_k(j-1) < G_k(m) < G_k(j)$, which, due to the strictly increasing nature of G_k, requires $j - 1 < m < j$, an impossibility since m and j belong to P. This contradiction shows that $S = \varnothing$ and $G_k(i) \leq G_{k+1}(i)$ for each $i \in P$.

Next, we so define the sequence f that for each $k \in P$, $f_k = G_k(k) \in P$. We assert that f increases strictly. For if $k \in P$, then

$$f_k = G_k(k) \leq G_{k+1}(k) < G_{k+1}(k+1) = f_{k+1}$$

by virtue of our observations above and the fact that G_{k+1} is strictly increasing. Thus $B = A \circ f$ is a subsequence of A. Now for any $k \in P$, $G_k(k) \in T_k$, and consequently for each such k,

$$0 \leq |M - B_k| = |M - A(f(k))| < 1/k$$
$$(0 \leq |m - B_k| = |m - A(f(k))| < 1/k).$$

Application of Lemma 5.14 and Theorem 5.13 now yields the desired conclusion.

6.7   Theorem.   If A is a sequence with values in Re and $\limsup\limits_{n} A_n$ (liminf A_n) is $+\infty$ or $-\infty$, then there exists a subsequence of A whose limit is $\limsup\limits_{n} A_n$ ($\liminf\limits_{n} A_n$).

Proof.   If $\limsup\limits_{n} A_n = -\infty$ ($\liminf\limits_{n} A_n = +\infty$) then by virtue of the fact that $\liminf\limits_{n} A_n \leq \limsup\limits_{n} A_n$ and Theorem 5.26, it follows that the sequence A has as its limit $\limsup\limits_{n} A_n$ ($\liminf\limits_{n} A_n$) and there is nothing to prove. Accordingly we assume $\limsup\limits_{n} A_n = +\infty$ ($\liminf\limits_{n} A_n = -\infty$).

We so define the sequence T of subsets of P that for each $k \in P$,

$$T_k = \{n \mid n \in P, n \geq k, \text{ and } A_n > k\}$$
$$(T_k = \{n \mid n \in P, n \geq k, \text{ and } A_n < -k\}).$$

We see that if $k \in P$, $k' \in P$, $k \leq k'$, and $n \in T_{k'}$, then $n \geq k' \geq k$, $A_n > k' \geq k$ ($A_n < -k' \leq -k$), and consequently $n \in T_k$, whence $T_{k'} \subset T_k$.

Also, T_k is an infinite set for each $k \in P$. To prove this, we take an arbitrary positive integer L. In the notation of Definition 5.25, it follows that there exists $N \in P$ such that

(1) $$\sup \mathscr{A}_n > k, \qquad (\inf \mathscr{A}_n < -k)$$

whenever $N < n \in P$. We take $q = L + N + k$, and using Theorem 5.5 we note that there exists $x \in \mathscr{A}_q$ such that

(2) $$k < x \leq \sup \mathscr{A}_q, \qquad (\inf \mathscr{A}_q \leq x < -k).$$

Because of the way $\mathscr{A}_q$ is defined, there exists $p \in P$ such that $p \geq q$ and $x = A_p$. Since $p \geq q > L$ and $p \geq k$, we conclude that $p \in T_k$ and that T_k is unbounded above due to the arbitrary nature of L. Thus by Corollary 3.17, T_k is infinite.

At this point we define a sequence of functions G exactly as in Theorem 6.6, as well as a function f, also exactly as in that theorem. It follows with no changes in the proof there given that f is a strictly increasing sequence with $f_k \in T_k$ for each $k \in P$, and therefore the function $B = A \circ f$ is such a subsequence of A that

$$B_k = A(f(k)) > k, \qquad (B_k = A(f(k)) < -k)$$

for each $k \in P$. Application of Lemma 5.14 and Theorem 5.13 now yields the desired result.

Exercises

1. Assume $x \in R$. Prove the existence of a unique function f so defined on P that $f(1) = x$ and $f(n + 1) = x \cdot f(n)$ for each $n \in P$. The resulting function is the one whose value $f(n)$ is conventionally denoted by x^n. Using proof by induction, show that $x^m \cdot x^n = x^{m+n}$ and $(x^m)^n = x^{mn}$ whenever m and n belong to P.

2. Prove that there exists a unique function f so defined on P that $f(1) = 1$ and $f(n + 1) = (n + 1) \cdot f(n)$ whenever $n \in P$. This function is the one whose value $f(n)$ is usually denoted by $n!$.

3. Show that there exists a unique function f so defined on P that $f(1) = 1$ and $f(n + 1) = f(n) + 2n + 1$ whenever $n \in P$. Show that if $g(n) = n^2$ for each $n \in P$, then g possesses the properties required of f; hence $f = g$. In this case we have found an explicit form of the required function.

4. Assume that A is a (finite) real-valued sequence. Show that there exist functions f and g such that $f(1) = A_1 = g(1)$, $f(n + 1) = f(n) + A_{n+1}$ and $g(n + 1) = A_{n+1} \cdot g(n)$ whenever $n \in P$. The two functions are the ones whose values $f(n)$ and $g(n)$ are frequently denoted by $\sum_{i=1}^{n} A_i$ and $\prod_{i=1}^{n} A_i$, respectively. In both these expressions, i plays the role of a dummy variable.

   Prove that if $c \in R$, $D = cA$, and $n \in P$, then $\sum_{i=1}^{n} D_i = c \sum_{i=1}^{n} A_i$; if A and B are (finite) real-valued sequences and $D = A + B$, then $\sum_{i=1}^{n} D_i = \sum_{i=1}^{n} A_i + \sum_{i=1}^{n} B_i$ whenever $n \in P$.

7

Functions of a Real Variable; Limits and Continuity

We have dealt with functions in general, and we have given a rather thorough treatment to numerical-valued sequences in particular. We shall now turn to the consideration of functions whose domains are more general subsets of Re than P, and whose values are in Re. These will be called *functions of a real variable*. As we use this term, it includes sequences whose values are in Re, although when we think of functions of a real variable, we usually have in mind functions whose domains are not *discrete* subsets of Re, that is, we think of domains in which the members may be arbitrarily near one another, unlike the members of P. For instance, we might imagine functions defined and real valued on the entire real number system, or on the set of all rational numbers. A type of domain commonly associated with such functions is the *interval*, of which there are several kinds. We will define them later in this chapter.

We start by defining for functions of a real variable a concept of *limit* that is a generalization of Definition 5.9. We shall discuss its intuitive significance a little later.

7.1 **Definition.** If f is a function whose domain and range are both subsets of Re, $A \in Re$, $a \in Re$, and a is a cluster point of $\mathscr{D}(f)$ (recall Definitions 3.21, 3.22, 5.6, and 5.7), then we say that A *is a limit of* f *at* a if and only if for each neighborhood I of A, there exists a neighborhood J of a such that $f(x) \in I$ whenever $x \in \mathscr{D}(f)$ and $a \neq x \in J$.

Because neighborhoods of finite and infinite numbers have somewhat different forms, the definition just given takes different forms depending on whether a and A are finite or not. The various possible cases have their corresponding expressions as follows:

CASE 1. $a \in R$, $A \in R$. Given any positive real number ϵ, there exists $\delta > 0$ such that $A - \epsilon < f(x) < A + \epsilon$, that is, $|f(x) - A| < \epsilon$, whenever $a \neq x \in \mathscr{D}(f)$ and $a - \delta < x < a + \delta$, that is, $x \in \mathscr{D}(f)$ and $0 < |x - a| < \delta$.

CASE 2. $a \in R$, $A = +\infty\,(-\infty)$. Given any finite positive number M, there exists $\delta > 0$ such that $f(x) > M\,(f(x) < -M)$ whenever $a \neq x \in \mathscr{D}(f)$ and $a - \delta < x < a + \delta$, that is, $x \in \mathscr{D}(f)$ and $0 < |x - a| < \delta$.

CASE 3. $a = +\infty \, (-\infty)$, $A \in R$. Given any positive number ϵ, there exists a finite positive number K such that $A - \epsilon < f(x) < A + \epsilon$, that is, $|f(x) - A| < \epsilon$, whenever $x \in \mathscr{D}(f)$ and $+\infty > x > K\,(-\infty < x < -K)$.

CASE 4. $a = +\infty$ (or $a = -\infty$) and $A = +\infty$ (or $A = -\infty$). Given any finite positive number M, there exists a finite positive number K such that $f(x) > M$ (or $f(x) < -M$) whenever $x \in \mathscr{D}(f)$ and $+\infty > x > K$ (or $-\infty < x < -K$).

A glance at Cases 3 and 4 will convince the reader that if a sequence f has a limit A in the sense of Definition 5.9, then it has A as its limit in the sense of Definition 7.1, since $+\infty$ is a cluster point of $P = \mathscr{D}(f)$. Thus Definition 7.1 is a generalization of Definition 5.9.

7.2 Definition. If f is a function, then we shall say that *f has a limit at a* if and only if $\mathscr{D}(f) \subset Re$, $\mathscr{R}(f) \subset Re$, a is a cluster point of $\mathscr{D}(f)$, and there exists $A \in Re$ such that A is a limit of f at a.

From this definition, we have no immediate assurance that if f has a limit at a, then the limit is unique. In the special case that f is a sequence, we know this is so, but we want to show it is true in general. In fact, the reason for demanding that a be a cluster point of $\mathscr{D}(f)$ is to ensure uniqueness. If a were not a cluster point of $\mathscr{D}(f)$, then there would exist, by Theorems 3.23 and 5.8, a neighborhood I of a such that

$$\mathscr{D}(f) \cap (I - \langle a \rangle) = \varnothing ,$$

in which case we see that for any $A \in Re$, the condition for A to be a limit of f would be satisfied vacuously. Thus we need to require at least that a be a cluster point of $\mathscr{D}(f)$ in order to ensure uniqueness of the limit; and as we shall soon show, that is all we need.

The proof we are about to give of this fact, together with the proofs of numerous other properties of functional limits that we shall give later, will be seen to depend heavily on the results we have already established for sequences. Actually, they could all be proved without reference to sequences, and there would be pedagogical value in such a procedure. To acquaint the reader with the techniques involved, a few such theorems appear in the exercises, to be worked out directly from the concept of functional limits. However, we have taken great pains to do a thorough job on sequences, and it happens that when these sequential theorems are expressed in the language of neighborhoods, they yield very quickly the results we want on limits of functions of a real variable, sometimes covering all the possible cases of finite or infinite values at once, and so we use them for brevity. Also, in this way we emphasize the important connections between limits of functions of a real variable and those of certain sequences associated with those functions.

7.3   Lemma.   If $S \subset Re$ and a is a cluster point of S, then there exists a sequence X such that $\mathscr{R}(X) \subset (S - \langle a \rangle)$ and $\lim_n X_n = a$.

Proof.   We so define the sequence T of subsets of Re that for each $n \in P$,

$$T_n = S \cap \{x \mid a - 1/n < x < a + 1/n\} - \langle a \rangle \quad \text{if} \quad a \in R;$$
$$T_n = S \cap \{x \mid x > n\} - \langle a \rangle \quad \text{if} \quad a = +\infty;$$
$$T_n = S \cap \{x \mid x < -n\} - \langle a \rangle \quad \text{if} \quad a = -\infty.$$

Since a is a cluster point of S, it follows that no matter which of the three cases listed occurs, T_n is infinite and so is non-empty for each $n \in P$. We may therefore so define[1] the sequence X that for each $n \in P$, $X_n \in T_n$. Clearly $X_n \in (S - \langle a \rangle)$ for each $n \in P$; hence $\mathscr{R}(X) \subset (S - \langle a \rangle)$. Also, according to whether $a \in R$, $a = +\infty$, or $a = -\infty$, we see that for each $n \in P$,

$$0 < |X_n - a| < \frac{1}{n}, \quad +\infty > X_n > n, \quad \text{or} \quad -\infty < X_n < -n.$$

respectively, whence by Lemma 5.14 and Theorem 5.13, we have in any case $\lim_n X_n = a$.

7.4   Lemma.   If f is a function, $\mathscr{D}(f) \subset Re$, $\mathscr{R}(f) \subset Re$, $a \in Re$, $A \in Re$ a is a cluster point of $\mathscr{D}(f)$, and A is a limit of f at a, then for every sequence X such that $\mathscr{R}(X) \subset (\mathscr{D}(f)) - \langle a \rangle)$ and $\lim_n X_n = a$, we have $\lim_n f(X_n) = A$.

Proof.   We take an arbitrary neighborhood I of A. Then, by Definition 7.1, there exists a neighborhood J of a such that

(1) $$f(x) \in I \quad \text{whenever} \quad a \neq x \in J.$$

Now if X is any sequence satisfying the hypotheses of our lemma, then by Definition 5.9 there exists a positive integer N such that $X_n \in J$ whenever $N < n \in P$; thus $f(X_n) \in I$ for each such n, due to (1). But this is the condition that the sequence $\{f(X_n)\}$ have A as its limit, and the proof is complete.

7.5   Corollary.   If f is a function, $\mathscr{D}(f) \subset Re$, $\mathscr{R}(f) \subset Re$, $a \in Re$, a is a cluster point of $\mathscr{D}(f)$, A and A' are limits of f at a, then $A = A'$.

Proof.   By Lemma 7.4, if X is any sequence with $\mathscr{R}(X) \subset (\mathscr{D}(f) - \langle a \rangle)$ and $\lim_n X_n = a$, then $A = \lim_n f(X_n)$ and $A' = \lim_n f(X_n)$. Since we know by Theorem 5.11 that sequential limits are unique, then this last implies

---

[1] Here we are using a form of the axiom of choice. Recall the closing remarks in Chapter 1.

$A = A'$. By Lemma 7.3, there necessarily exists a sequence X with the properties just listed; hence we can conclude that $A = A'$.[2]

The uniqueness of functional limits, which this last result guarantees, justifies the following definition.

7.6 Definition. If f is a function having a limit at a, we say that *the limit of f at a exists*, and we agree to denote by $\lim_a f$ the (unique) member of *Re* that is its limit at a.

As was the case in sequential limits, we find it convenient to introduce an alternative notation for $\lim_a f$, namely $\lim_{x \to a} f(x)$ [read: *the limit of $f(x)$ as x approaches* (or *tends to*) *a*]. In this expression, x plays the role of a dummy variable, and may be replaced by any other variable not occurring in the entire expression.

The chief advantage of this notation is that it eliminates the necessity of giving names to a great many functions, although it requires that the reader make a correct, though usually obvious, interpretation. For example, consider $\lim_{x \to 2} (x^2 + 2x)$ $\big($or, equivalently, $\lim_{y \to 2} (y^2 + 2y)\big)$. The reader has to interpret this to mean that we are concerned with the limit of the function f at 2 whose value $f(x)$ at an arbitrary $x \in \mathscr{D}(f)$ is $x^2 + 2x$. Somewhere the reader must be told what the domain of f is supposed to be. As it happens in this example, the value of the limit does not depend on what the domain may be, so long as it is a subset of *Re* for which 2 is a cluster point. However, it is easy to construct examples in which the domain plays a determining role in the evaluation of the limit. Some authors use the unwritten convention that in expressions such as

$$\lim_{x \to 1} (x^2 - 3x + 1) \quad \text{and} \quad \lim_{x \to -1} (x^2 - 1)/(x + 1),$$

where there is no clue given as to the domain of the function, the domain may be tacitly regarded as the set of all real numbers for which the expressions in question take values in *Re*. We may occasionally use this convention when there is little likelihood of confusion; otherwise we will be careful to specify the domain in question.

Before proceeding with the development of our general limit theory, we shall pause to take a close look at what is involved in our limit definition and work some specific examples. First, we note that our definition of $\lim_a f$ specifically excludes reference to the value of f at a. The existence of $\lim_a f$ is determined by the behavior of f near a, not at a. Also, a is required

---

[2] It should be clear how Lemma 7.3 plays an essential role here, for the conclusion $A = A'$ can be inferred only in case the initial statement in the chain of implications leading to it can be satisfied for some sequence X.

to be a cluster point of $\mathscr{D}(f)$ but, as we know, this does not mean necessarily that $a \in \mathscr{D}(f)$. It often happens that $a \notin \mathscr{D}(f)$, in which case f is not even defined at a, but this circumstance has no effect whatever on the existence or non-existence of $\lim_a f$.

Expressed rather loosely, our definition says that if x is taken close to a, but not at a, then the corresponding values $f(x)$ must all be close to A. As was the case in sequential limits, the definition neither prohibits nor demands that the function assume the limit value itself. Also, in order to determine that a particular number A shall be the limit of f at a, it is more or less necessary to guess the value of A beforehand. However, there is a criterion, analogous to the Cauchy criterion for sequences, that enables us to determine the existence of finite limits. Also, Lemma 7.4 and Corollary 7.5 suggest a useful method of proving the non-existence of the limit of f at a. All we need to do for this purpose is to discover a sequence X with $\mathscr{R}(X) \subset (\mathscr{D}(f) - \langle a \rangle)$ and $\lim_n X_n = a$ for which the sequence $\{f(X_n)\}$ has no limit, or else discover two such sequences X and X' for which $\lim_n f(X_n)$ and $\lim_n f(X'_n)$ exist but are unequal.

We shall give graphical or pictorial representations of our limit concept that may be helpful in making the idea comprehensible. We find it necessary to consider separately the four cases under Definition 7.1. The worked examples following each of these cases are for the purpose of familiarizing the reader with the techniques involved in this kind of analysis. However, we will prove some general theorems later that eliminate the necessity for such analysis in many cases. Also, it will be shown that limits of the kinds occurring under Cases 3 and 4 can be converted to the types of Cases 1 and 2 by making an appropriate change of variable.

7.7 Graphical Representation of Limits When $a \in R$, $A \in R$; **Worked Examples.** We assume that f is a given function; since we are going to plot it as in Figure 2, around the general vicinity of $x = a$, it is convenient to assume it is defined for all real x near a, and finite there. We may ignore completely the value of f at a, if indeed it has a value there. We suppose that ϵ is an arbitrary positive number. Now, assuming that $A = \lim_a f$, we draw the horizontal lines $y = A + \epsilon$ and $y = A - \epsilon$. Then it must be possible to show the existence of a neighborhood of a, and correspondingly a positive number δ, such that if we draw the vertical lines $x = a - \delta$ and $x = a + \delta$, then the portion of the graph representing f between these two lines lies entirely between the lines $x = A + \epsilon$ and $y = A - \epsilon$, with the possible exception of the point on the graph corresponding to $x = a$

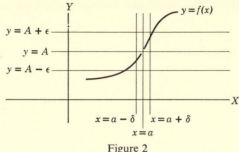

Figure 2

itself. Incidentally, when proving the existence of $\delta > 0$ with the properties we desire it is not necessary to find the largest possible value for δ, if such exists; it is frequently much easier to find a smaller value for δ, and this is sufficient for our purposes. If, on the other hand, there exists some number $\epsilon > 0$ such that, no matter how small a positive number δ we take, there is always some point on the graph between $x = a - \delta$ and $x = a + \delta$ (apart from the point corresponding to $x = a$) that does not lie between the horizontal lines mentioned, then A is not the limit of f at a.

Again looking at Figure 2, we imagine what happens in case ϵ is decreased. The lines $y = A + \epsilon$ and $y = A - \epsilon$ are brought closer together, and it is apparent that it may be necessary to find correspondingly a smaller value of δ to satisfy the conditions of our limit definition. In brief, δ is usually going to depend on ϵ; and the smaller ϵ may be, the smaller we must expect δ to be.

We shall work two examples illustrating the technique involved in certain simple limit problems. We first try to determine $\lim\limits_{x \to 2} (x^2 + x)$, where here we shall assume the domain of the indicated function is R. We graph the function $f(x) = x^2 + x$ around $x = a = 2$, as in Figure 2, and decide by inspection that the limit A in question is probably 6. In accordance with Definition 7.1, we assume that ϵ is an arbitrary positive number. We must find $\delta > 0$ so that

(1) $|x^2 + x - 6| < \epsilon$ whenever $0 < |x - 2| < \delta$.

We observe that $x^2 + x - 6$ factors; in fact, $x^2 + x - 6 = (x + 3) \cdot (x - 2)$. We expect, in view of (1), that we will have to make $|x - 2|$ rather small; as a start, we may take $|x - 2| < 1$. Any positive number other than 1 would do, so long as the function does not have serious irregularities within the corresponding neighborhood of $x = 2$. This means we are restricting ourselves to values of x such that $1 < x < 3$. The reason for such an initial restriction is to control the factor $x + 3$. We see that for any x satisfying $1 < x < 3$, we have $|x + 3| < 6$. Consequently, for each such x, we have

(2) $$|x^2 + x - 6| = |x + 3| \cdot |x - 2| < 6 |x - 2|.$$

Now it is apparent that the left term of the relation (2) will be less than ϵ if the right term is less than ϵ; this can clearly be achieved if we take $|x - 2| < \epsilon/6$. Thus we conclude that (1) holds provided we take for δ any positive number that is not greater than 1 and $\epsilon/6$ simultaneously.[3] We have to require that δ be not greater than 1 to ensure that (2) holds and not greater than $\epsilon/6$ to complete the job. We may think of the number 1 as a suitable value for δ whenever ϵ is fairly large, that is $\epsilon \geq 6$; and $\epsilon/6$ as a suitable value for δ whenever ϵ is rather small, that is, $\epsilon < 6$. Incidentally, it may be noted that in this example, the relation (1) holds even when $x = 2$.

We consider a second example, $\lim\limits_{x \to -1} (x^3 + 1)/(x + 1)$, where the indicated function is defined for all real numbers except $x = -1$. We graph the function $f(x) = (x^3 + 1)/(x + 1)$ around the point $a = -1$, much as was done in Figure 2. From our graph we decide that the limit is probably 3. We assume that ϵ is an arbitrary positive number. We must find $\delta > 0$ so that

(3) $$\left| \frac{(x^3 + 1)}{(x + 1)} - 3 \right| < \epsilon \qquad \text{whenever } 0 < |x + 1| < \delta$$

Now, provided $x \neq -1$, we see that

$$\left| \frac{x^3 + 1}{x + 1} - 3 \right| = |(x^2 - x + 1) - 3| = |x^2 - x - 2|;$$

hence for all real $x \neq -1$,

(4) $$\left| \frac{x^3 + 1}{x + 1} - 3 \right| = |x^2 - x - 2| = |x - 2| \cdot |x + 1|$$

As in the first example, we initially restrict x so that $0 < |x + 1| < 1$, that is, $-2 < x < 0$, $x \neq -1$. For all such values of x, we see that $|x - 2| < 4$. Thus from (4) we have, for each such x,

(5) $$\left| \frac{x^3 + 1}{x + 1} - 3 \right| = |x - 2| \cdot |x + 1| < 4 |x + 1|$$

---

[3] If one wishes to give an explicit value for δ that is bound to be smaller than both 1 and $\epsilon/6$, one can take $\delta = \epsilon/(6 + \epsilon)$ (cf. Problem 59 in the Exercises at the end of Chapter 2). Otherwise, the smaller of the two numbers 1 and $\epsilon/6$ can be taken for δ.

To make the left side of (5) less than ϵ, we need only make the right side less than ϵ. This will be so provided $|x + 1| < \epsilon/4$. Thus we take δ to be any number not greater than 1 and $\epsilon/4$ simultaneously. In particular, we may take δ to be the smaller of these two numbers, or, alternatively, we may take $\delta = \epsilon/(4 + \epsilon)$. Now we see that (3) holds for our choice of δ, and so we have proved that $\lim\limits_{x \to -1} (x^3 + 1)/(x + 1) = 3$.

7.8 Graphical Representation of Limits When $a \in R$ **and** $A = +\infty(-\infty)$; **Worked Examples.** For convenience we will picture only the case $A = +\infty$; the corresponding picture for $A = -\infty$ is obtained effectively by rotating the graph shown in Figure 3 about the X-axis. We show thereon a portion of the graph of a typical function f around the vicinity of $x = a$. It has been assumed for convenience that f is defined for all real numbers near a, but we have specifically ignored the value of f at a, if indeed it has a value there. We assume that M is an arbitrary finite positive number. We draw the line $y = M$. If $A = +\infty$ is the limit of f at a, then there must exist a neighborhood of a, and a corresponding number $\delta > 0$, such that if we draw the vertical lines $x = a - \delta$ and $x = a + \delta$, then the portion of the graph between these lines lies entirely above the line $y = M$, with the possible exception of the point on the graph corresponding to $x = a$ itself. As in Case 1, it is necessary only to find some $\delta > 0$ satisfying our definition, not the largest possible value of δ, in case such exists. If, on the other hand, there exists some finite positive number M for which, no matter how small a value for δ we take, there is always some point on the graph between the vertical lines $x = a - \delta$ and $x = a + \delta$ with x-coordinate different from a that does not lie above the line $y = M$, then $A = +\infty$ is not the limit of f at a.

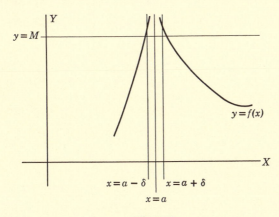

Figure 3

We see from Figure 3 that if M is increased, then δ may have to be decreased to achieve the desired end. Thus we expect that, in general, δ will depend upon M and will decrease as M increases.

We consider the function f so defined that $f(x) = 1/(x-1)^2$ for all real $x \neq 1$, and we will prove that $\lim\limits_{x \to 1} f(x) = +\infty$. We graph a portion of the function f around $a = 1$, as in Figure 3. According to Definition 7.1, Case 2, we must find $\delta > 0$ so that

(1) $$\frac{1}{(x-1)^2} > M \qquad \text{whenever } 0 < |x-1| < \delta.$$

Now the inequality in (1) is equivalent to $0 < (x-1)^2 < 1/M$, that is, to $0 < |x-1| < 1/\sqrt{M}$. Thus if we take $\delta = 1/\sqrt{M}$, the desired condition (1) holds.

Next, we will prove that if $f(x) = (x^2 - 6)/(x+2)^2$ for all real $x \neq -2$, then $\lim\limits_{x \to -2} f(x) = -\infty$. We graph this function around $a = -2$ in a manner similar to that of Figure 3. We assume M is an arbitrary finite positive number, and we have to find $\delta > 0$ so that

(2) $$\frac{x^2 - 6}{(x+2)^2} < -M \qquad \text{whenever } 0 < |x+2| < \delta.$$

We observe first that if we restrict x so that $-1\frac{3}{4} < x < -2\frac{1}{4}$, that is, $|x+2| < \frac{1}{4}$, then for all such x, $x^2 < (2\frac{1}{4})^2 = 5\frac{1}{16}$, whence $x^2 - 6 < 5\frac{1}{16} - 6 = -\frac{15}{16} < -\frac{1}{4}$. Thus for all such $x \neq -2$, we have $(x^2 - 6)/(x+2)^2 < -1/4(x+2)^2$. Thus (2) will hold provided $-1/4(x+2)^2 < -M$, that is, if $M < 1/4(x+2)^2$, or equivalently, $0 < |x+2| < 1/2\sqrt{M}$. Thus if we take for δ any positive number that is not greater than $\frac{1}{4}$ and $1/2\sqrt{M}$ simultaneously, it will suffice to make the condition (2) hold; consequently, $\lim\limits_{x \to -2} f(x) = -\infty$.

7.9 Graphical Representation of Limits When $a = +\infty(-\infty)$ and $A \in R$; Worked Examples.

For convenience we show only the case $a = +\infty$. The case $a = -\infty$ can be obtained by rotating the graph in Figure 4 about the Y-axis. There we have shown a typical function f assumed defined for all $x \in Re$. We are concerned only with the values taken by f for large finite values of x. We assume ϵ is an arbitrary positive number, and assuming that $A = \lim\limits_{a} f$, we draw the horizontal lines $y = A + \epsilon$ and $y = A - \epsilon$. There must exist some finite positive number K such that the entire portion of the graph of f to the right of K, with the possible exception of the point on it corresponding to $x = +\infty$, lies between the

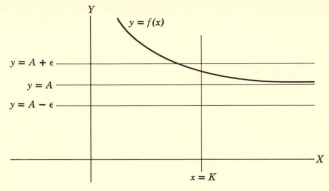

Figure 4

lines $y = A + \epsilon$ and $y = A - \epsilon$. If there exists some $\epsilon > 0$ such that no matter how large a value for K we take there exists some point on the graph to the right of K that does not lie between the lines $y = A + \epsilon$ and $y = A - \epsilon$ (with the possible exception of the point corresponding to $x = +\infty$), then A is not the limit of f at A.

If ϵ is imagined to decrease, then the lines $y = A + \epsilon$ and $y = A - \epsilon$ draw closer together, and it appears that K may have to be increased to achieve the desired end. Thus we expect that as ϵ decreases, K must increase, in general.

We consider the example $f(x) = (x + 1)/x$ for all real $x \neq 0$. We shall show that $\lim\limits_{x \to +\infty} f(x) = 1$. We plot the function f much as was done in Figure 4. We take an arbitrary positive number ϵ. We must find a finite positive number K such that

(1) $|f(x) - 1| < \epsilon$ whenever $K < x < +\infty$.

Now $f(x) - 1 = 1/x$ whenever $0 \neq x \in R$, as we see from direct calculation. The relation $|f(x) - 1| = |1/x| < \epsilon$ is equivalent to $|x| > 1/\epsilon$. Thus if we take $K = 1/\epsilon$, the desired relation (1) will hold, whence $\lim\limits_{x \to +\infty} f(x) = 1$. It may likewise be noted that the relation

(2) $|f(x) - 1| < \epsilon$ whenever $-\infty < x < -K$

is valid if $K = 1/\epsilon$, whence $\lim\limits_{x \to -\infty} f(x) = 1$, also.

Next we consider the function f so defined that $f(x) = (2x^2 + x - 1)/(x^2 + 5x + 2)$ for all real x not roots of the denominator. We shall prove that $\lim\limits_{x \to +\infty} f(x) = 2$. We graph f as in Figure 4. We assume that ϵ is an arbitrary positive number. We have to determine K so that

(3) $|f(x) - 2| = \dfrac{|-9x - 5|}{|x^2 + 5x + 2|} < \epsilon$ whenever $K < x < +\infty$.

Now, dividing through the numerator and denominator of this last expression by x^2, we obtain, for all $x > 0$,

(4)
$$\left| \frac{9x + 5}{x^2 + 5x + 2} \right| = \frac{\left| \dfrac{9}{x} + \dfrac{5}{x^2} \right|}{\left| 1 + \dfrac{5}{x} + \dfrac{2}{x^2} \right|}$$

If we make the restriction that $x \geq 1$, then $0 < (1/x) \leq 1$, and (4) yields

(5)
$$\frac{|9x + 5|}{|x^2 + 5x + 2|} < \frac{9}{|x|} + \frac{5}{|x|^2} < \frac{9}{|x|} + \frac{5}{|x|} = \frac{14}{|x|}.$$

If we now also demand that $x > 14/\epsilon$, then from (5) we obtain

$$\left| \frac{9x + 5}{x^2 + 5x + 2} \right| < \frac{14}{|x|} < \epsilon.$$

Thus if we take for K any finite number that is not less than 1 and $14/\epsilon$ simultaneously, then (1) holds as required, and so $\lim\limits_{x \to +\infty} f(x) = 2$.

Incidentally, with some modifications in the analysis given above, it is possible to show that $\lim\limits_{x \to -\infty} f(x) = 2$.

7.10   Graphical Representation of Limits When $a = +\infty$ **or** $a = -\infty$ **and** $A = +\infty$ **or** $A = -\infty$; **Worked Examples.**   For convenience we show only the case $a = +\infty$ and $A = +\infty$. The other three cases can be obtained by rotating the graph shown in Figure 5 about the X- or Y-axes,

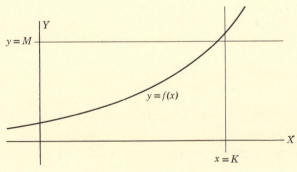

Figure 5

or both. The function shown there is a typical one, assumed to be defined for all large positive values of x. We assume that M is an arbitrary finite positive number, and we draw the line $y = M$. If $A = +\infty = \lim_{+\infty} f$, then there must exist some finite positive number K such that for all x, $K < x < +\infty$, the corresponding portion of the graph lies above the line $y = M$. If there exists some M such that no matter how large a value of K we take there is always some point on the graph to the right of the line $x = K$ that does not lie above $y = M$, except possibly the point corresponding to $x = +\infty$, then the limit of f at $+\infty$ is not $+\infty$.

It is also clear from Figure 5 that if M increases, then generally K must likewise increase.

We consider as an example $f(x) = x^2 - 10$ for all $x \in R$. The graph of this function may be drawn as in Figure 5. We shall show that $\lim_{x \to +\infty} f(x) = +\infty$. We let M denote an arbitrary finite positive number, and we must find K such that $0 < K < +\infty$ and

(1) $\qquad f(x) = x^2 - 10 > M \qquad$ whenever $K < x < +\infty$.

First we restrict x so that $x > 1$; then $x^2 - 10 > x - 10$. If we take $x > 1$ and $x > M + 10$ simultaneously, then we will have $f(x) = x^2 - 10 > x - 10 > M$. Hence if we take K to be any finite number that is simultaneously not less than 1 and $M + 10$, then (1) will hold. Since M is positive, then we may take $K = M + 10$; for this value of K, the requirement $K \geq 1$ is automatically satisfied. Thus $\lim_{x \to +\infty} f(x) = +\infty$. By making a few modifications in the analysis just completed, it is easily shown that $\lim_{x \to -\infty} f(x) = +\infty$ also.

Next we consider $f(x) = x^3/(10x^2 + 1)$ for all real x. The graph of f should be drawn similar to the graph in Figure 5. We will prove that $\lim_{x \to -\infty} f(x) = -\infty$. We assume that M is an arbitrary finite positive number, and we must find K such that $0 < K < +\infty$ and

(2) $\qquad f(x) = \dfrac{x^3}{10x^2 + 1} < -M \qquad$ whenever $-\infty < x < -K$.

We see that for all real $x < -1$,

(3) $\qquad \dfrac{x^3}{10x^2 + 1} = \dfrac{x}{10 + 1/x^2} < \dfrac{x}{11} .$

Hence if we take $x < -1$ and $x < -11 M$ simultaneously, then from (3) we will have

$$\frac{x^3}{10x^2 + 1} < \frac{x}{11} < -M.$$

Thus (2) holds if we take for K any finite positive number that is not less than 1 and $11M$ simultaneously. In particular, we could take $K = 1 + 11M$. By making a few modifications in the above it is easily shown that $\lim\limits_{x \to +\infty} f(x) = +\infty$.

In Lemma 7.4 we showed a connection between the limit of a function and the limits of certain sequences of its values near a. We will now prove the converse of that lemma, thereby establishing that the property in question is a criterion for the existence of the functional limit at a. We prove it separately, mainly to isolate the technique employed, which occurs in various forms throughout analysis. Before doing so, we need two additional results that are of interest in themselves.

7.11   Lemma. If $a \in Re$, I and J are neighborhoods of a, then $I \cap J$ is a neighborhood of a.

Proof. According to whether $a \in R$, or $a = +\infty$, or $a = -\infty$, there exist finite positive numbers s and t such that

$$I = \{x \mid |x - a| < s\} \quad \text{and} \quad J = \{x \mid |x - a| < t\}$$

or $\qquad I = \{x \mid x > s\} \quad \text{and} \quad J = \{x \mid x > t\}$

or $\qquad I = \{x \mid x < -s\} \quad \text{and} \quad J = \{x \mid x < -t\}$, respectively.[4]

In case $a \in R$, we may let u denote the smaller of the numbers s and t; then $0 < u < +\infty$. Now if x is any real number such that $|x - a| < u$, then $|x - a| < s$ and $|x - a| < t$ both hold, and conversely. From this we infer that $I \cap J = \{x \mid |x - a| < u\}$; thus $I \cap J$ is a neighborhood of a.

In case $a = +\infty \, (-\infty)$, we may let v denote the larger of the numbers s and t; then $0 < v < +\infty$. Also, if x is any real number such that $x > v \, (x < -v)$, then certainly $x > s$ and $x > t \, (x < -s \text{ and } x < -t)$

---

[4] It has long been customary in mathematics to use the variables δ and ϵ in defining finite functional limits at finite real points. We have followed this practice. Since we usually think of neighborhoods in connection with limits as being small, people have come to associate smallness with variables such as δ and ϵ. There is no logical reason for this; variables in mathematics do not acquire properties by virtue of their names, nor do the properties of a variable necessitate the use of any particular name for it. Similarly, one might have the feeling that M and K as used in our definition of infinite limits or limits at infinity suggest largeness. It is true that we have in mind large values when dealing with such limits, but any two names at all could be used for the variables involved in the situation. We could just as well use δ and ϵ in these cases, but in deference to the long-standing custom regarding their use, we did not. However, in the present proof, both finite and infinite situations occur, and in order to avoid confusion as well as the unnecessary introduction of four different names for our variables, we have introduced the new variables s and t for both purposes.

both hold, and conversely. Thus $I \cap J = \{x \mid x > v\}$ ($I \cap J = \{x \mid x < -v\}$), so that $I \cap J$ is a neighborhood of $+\infty$ ($-\infty$).

7.12   Corollary. If $S \subset Re$, a is a cluster point of S, and J is a neighborhood of a, then a is a cluster point of $S \cap J$.

Proof. We consider an arbitrary neighborhood I of a. Due to Lemma 7.11, $I \cap J$ is a neighborhood of a, whence $(I \cap J) \cap S = I \cap (J \cap S)$ is an infinite set by Definitions 3.22 and 5.7. But by these same definitions and the arbitrary nature of I, it follows that a is a cluster point of $J \cap S$.

7.13   Lemma. If f is a function whose domain and range are subsets of Re, $a \in Re$, $A \in Re$, U is a neighborhood of a and $\lim_n f(X_n) = A$ whenever X is a sequence such that $\lim_n X_n = a$ and $\mathscr{R}(X) \subset (U \cap \mathscr{D}(f) - \langle a \rangle)$, then $\lim_{x \to a} f(x) = A$.

Proof. We find it necessary for the first time to examine the logical structure of the negation of the statement "$\lim_{x \to a} f(x) = A$." Referring back to Definition 7.1, we see that this means there exists a neighborhood I of A for which there exists no neighborhood J of a such that $f(x) \in I$ whenever $a \neq x \in J \cap \mathscr{D}(f)$. In other words, no matter what neighborhood J of a may be given, there exists x such that $a \neq x \in J \cap \mathscr{D}(f)$ and $f(x) \notin I$.

We define the sequences T and S so that for each $n \in P$,

$$(1) \quad \left\{ \begin{aligned} &T_n = U \cap \{x \mid |x - a| < 1/n\}, \\ &\text{or } T_n = U \cap \{x \mid x > n\}, \\ &\text{or } T_n = U \cap \{x \mid x < -n\}, \end{aligned} \right.$$

according to whether $a \in R$, or $a = +\infty$, or $a = -\infty$, respectively, and

$$S_n = (T_n \cap \mathscr{D}(f) - \langle a \rangle) \cap \{x \mid f(x) \notin I\}.$$

If $n \in P$, then by virtue of Lemma 7.11, T_n is a neighborhood of a, and hence from the nature of I it follows that $S_n \neq \varnothing$. Thus there exists a sequence Y such that $Y_n \in S_n$, that is, $a \neq Y_n \in T_n \cap \mathscr{D}(f)$ and $f(Y_n) \notin I$, for each $n \in P$.[5] From (1) it now follows that $\lim_n Y_n = a$ no matter whether $a \in R$, $a = +\infty$, or $a = -\infty$. Also it is clear that $\mathscr{R}(Y) \subset (U \cap \mathscr{D}(f) - \langle a \rangle)$. On the other hand, for each $n \in P$, $f(Y_n) \notin I$, and so by Definition 5.9, the sequence $\{f(Y_n)\}$ does not have A as its limit. Consequently, if $\lim_n f(X_n)$ must equal A for every sequence X whose limit is a

---

[5] Here again we are using a form of the axiom of choice discussed at the end of Chapter 1.

and whose range is contained in $U \cap \mathscr{D}(f) - \langle a \rangle$, as our hypotheses require, then we must have $\lim_{x \to a} f(x) = A$, and the proof is complete.

7.14 Local Convention.   To simplify the statements of our theorems on limits, we shall henceforth frequently not trouble to specify that the functions occurring in them have domains and ranges in *Re*, nor that certain points must be cluster points of their domains. Since our limit concept was defined only for such functions, we shall agree that mention of such limits carries with it the tacit assumption that the functions and points concerned have the appropriate properties.

7.15 Lemma.   If $\lim_a f = A \in Re$, $T \subset Re$, and a is cluster point of $T \cap \mathscr{D}(f)$, then $\lim_a (f \mid T) = A$.

Proof.   According to Definition 7.1, if I is an arbitrary neighborhood of A, then there exists a neighborhood J of a such that $f(x) \in I$ whenever $a \neq x \in J \cap \mathscr{D}(f)$. Since $\mathscr{D}(f \mid T) = T \cap \mathscr{D}(f)$, then $(f \mid T)(x) = f(x) \in I$ whenever $a \neq x \in J \cap \mathscr{D}(f \mid T)$; consequently $\lim_a (f \mid T) = A$.

7.16 Corollary.   If f is a function, a is a cluster point of $\mathscr{D}(f)$ and J is a neighborhood of a, then $\lim_a f$ and $\lim_a (f \mid J)$ either both exist and are equal, or both fail to exist.

Proof.   Either both limits fail to exist, which is one of the possible conclusions, or else at least one of them exists. In case $\lim_a f \in Re$, then since a is a cluster point of $\mathscr{D}(f)$, it is also a cluster point of $J \cap \mathscr{D}(f) = T$ by Corollary 7.12; since $(f \mid T) = (f \mid J)$, then Lemma 7.15 applies and we conclude that $\lim_a (f \mid J) = \lim_a f \in Re$.

If $A = \lim_a (f \mid J) \in Re$, we may take an arbitrary neighborhood I of A and see that there exists a neighborhood K of a such that $f(x) = (f \mid J)(x) \in I$ whenever $a \neq x \in (K \cap J) \cap \mathscr{D}(f)$. Since $K \cap J$ is a neighborhood of a by Lemma 7.11, then $\lim_a f = A$, as required.

Lemma 7.15 shows that when a function is restricted (cf. Definition 1.41) and a is a cluster point of the domain of the restricted function, then we may evaluate the limit of the restricted function immediately, if we know the limit of the parent function at a. On the other hand, we cannot infer that the parent function has a limit at a simply because one of its restrictions has a limit there. For instance, let A be the set of all rational numbers and B the set of all irrational numbers, and define f on R so that $f(x) = 0$ if $x \in A$, $f(x) = 1$ if $x \in B$. From Definition 7.1 it follows readily that $\lim_0 (f \mid A) = 0$, $\lim_0 (f \mid B) = 1$, and with the help of Lemma 7.4

that f itself has no limit at 0 (cf. Problem 12 of the Exercises at the end of this chapter). Since a parent function may be regarded as a continuation (cf. Definition 1.38) of any of its restrictions, we may turn our point of view around and conclude that if a function has a limit at a given point, then generally we can make no assertion about the existence or non-existence of a limit at that point for any of its non-trivial continuations. Some of the most interesting problems of analysis have to do with finding continuations of certain functions in such a way as to preserve particular limit properties.

Corollary 7.16 shows that there do exist circumstances under which the existence of the limit of a suitable restriction of f at a allows us to infer that the limit of f itself at a exists, and also tells us its value. There is another situation of some importance which allows us to draw the same conclusion, and we take it up now.

7.17   Right- and Left-Hand Limits.   If $S \subset Re$ and $a \in Re$, then we define

$$S_a^+ = \{x \mid x \in S \text{ and } a < x\},$$
$$S_a^- = \{x \mid x \in S \text{ and } x < a\}.$$

Clearly $S_a^+ \cap S_a^- = \varnothing$ and $S_a^+ \cup S_a^- = S - \langle a \rangle$.

Now we consider a function f whose domain is $S \subset Re$ with $\mathscr{R}(f) \subset Re$ and we take its *right-* and *left-hand restrictions* $(f \mid S_a^+)$ and $(f \mid S_a^-)$, respectively. If we plot f in the usual fashion on a graph, it becomes apparent that these two restrictions correspond to the portions of the graph of f to the right and to the left of a, respectively. In case a is a cluster point of S_a^+ we may consider $\lim_a (f \mid S_a^+)$ if it exists; similarly we may consider $\lim_a (f \mid S_a^-)$ if a is a cluster point of S_a^- and this limit exists.

These limits are known as the *right-* and *left-hand limits of f at a*, and are denoted by $\lim_{x \to a+} f(x)$ and $\lim_{x \to a-} f(x)$, respectively.

7.18   Theorem.   If $a \in R$ and f is such a function with $\mathscr{D}(f) = S$ that $\lim_{x \to a+} f(x) = \lim_{x \to a-} f(x) \in Re$, then $\lim_{x \to a} f(x)$ exists and coincides with the common value of the right- and left-hand limits of f at a.

Proof.   We let $A = \lim_{x \to a+} f(x)$, and we take an arbitrary neighborhood I of A. By Definition 7.1, there exist neighborhoods J' and J'' of a such that

$$f(x) = (f \mid S_a^+)(x) \in I \qquad \text{whenever } a \neq x \in J' \cap S_a^+$$

and

$$f(x) = (f \mid S_a^-)(x) \in I \qquad \text{whenever } a \neq x \in J'' \cap S_a^-.$$

We let $J = J' \cap J''$; J is a neighborhood of a by Lemma 7.11. Also,

$$J \cap S - \langle a \rangle = (J' \cap J'') \cap (S_a^+ \cup S_a^-) \subset (J' \cap S_a^+) \cup (J'' \cap S_a^-)$$

Thus whenever $a \neq x \in J \cap S$ it follows that either $a \neq x \in J' \cap S_a^+$ or $a \neq x \in J'' \cap S_a^-$, whence from above $f(x) \in I$, and so $\lim_{x \to a} f(x) = A$.

7.19   Local Convention.   In Chapters 4 and 5 we proved a number of theorems about the limits of sums, products, etc., of two or more sequences (see, for example, Theorems 5.13, 5.16, 5.17, 5.18, 5.20 and some of their corollaries). We wish to prove analogous results for functions of a real variable. In the special case that these happen to be sequences with values in Re, we use the conventions established in § 5.10 for sums, products, etc. of such functions. If f and g are non-sequential functions with domains and ranges contained in Re, and if $k \in Re$, then we shall agree that $|f|, f + g, f - g, kf, fg$, and f/g are those functions whose values are

$$|f(x)|, \quad f(x) + g(x), \quad f(x) - g(x), \quad kf(x),$$
$$f(x)g(x), \quad \text{and} \quad \frac{f(x)}{g(x)},$$

respectively, for all $x \in Re$ such that the corresponding expressions are members of Re. Thus the domains of these functions are subsets of those of f and g and in some instances may be empty.

7.20   Theorem.   If f and g are functions with a common domain $S \subset Re$, $f(x) \leq g(x)$ whenever $x \in S$, $\lim_{x \to a} f(x) = A$ and $\lim_{x \to a} g(x) = B$, then $A \leq B$. Moreover, if $A = B$ and h is such a function with domain S that $f(x) \leq h(x) \leq g(x)$ whenever $x \in S$, then $\lim_{x \to a} h(x)$ exists and equals A.

Proof.   We take a neighborhood I of a. Note by virtue of Lemmas 7.3 and 7.4 that there exists a sequence X with $\lim_n X_n = a$, $\mathcal{R}(X) \subset (I \cap S - \langle a \rangle)$, $\lim_n f(X_n) = A$, and $\lim_n g(X_n) = B$. Since $f(X_n) \leq g(X_n)$ for each $n \in P$, the conclusion $A \leq B$ follows from Theorem 5.13.

To prove the second part of the theorem we consider an arbitrary sequence Y such that $\lim_n Y_n = a$ and $\mathcal{R}(Y) \subset (I \cap S - \langle a \rangle)$. Note that $f(Y_n) \leq h(Y_n) \leq g(Y_n)$ whenever $n \in P$. Use Lemma 7.4 and Theorem 5.13 to see that $\lim_n h(Y_n) = A$, and apply Lemma 7.13 to obtain $\lim_{x \to a} h(x) = A$.

7.21   Theorem.   If f is a function, $\mathcal{D}(f) = S \subset Re$, a is a cluster point of S, $k \in Re$, and $f(x) = k$ whenever $x \in S$, then $\lim_{x \to a} f(x) = k$.

Proof.   We take a neighborhood I of a and an arbitrary sequence X such that $\lim_n X_n = a$ and $\mathcal{R}(X) \subset (I \cap S - \langle a \rangle)$. Note that $\lim_n f(X_n) = k$, and apply Lemma 7.13.

7.22   Corollary.   If $c \in Re$, $\lim\limits_{x \to a} f(x) \in Re$, and $c \leq f(x)$ $(f(x) \leq c)$ whenever $a \neq x \in (\mathscr{D}(f) - \langle a \rangle)$, then

$$c \leq \lim_{x \to a} f(x) \qquad \left(\lim_{x \to a} f(x) \leq c \right)$$

Proof.   We define the function g in such a way that $g(x) = c$ whenever $x \in \mathscr{D}(f)$. Then $\mathscr{D}(g) = \mathscr{D}(f)$ and Theorems 7.20 and 7.21 may be applied.

7.23   Lemma.   If $\lim\limits_{x \to a} f(x) \in R$, then there exists a neighborhood I of a
such that $f(x) \in R$ whenever $a \neq x \in I \cap \mathscr{D}(f)$.
Proof.   We let $A = \lim\limits_{x \to a} f(x) \in R$. By Definition 7.1, there exists a neigh-
borhood I of a such that $A - 1 < f(x) < A + 1$ whenever $a \neq x \in I \cap \mathscr{D}(f)$; hence $f(x) \in R$ for each such x.

7.24   Lemma.   If $0 < A = \lim\limits_{x \to a} f(x) \in R$ $(0 > A = \lim\limits_{x \to a} f(x) \in R)$, then
there exists a neighborhood I of a such that

$$0 < (A/2) < f(x) < (3A/2) \qquad ((3A/2) < f(x) < (A/2))$$

whenever $a \neq x \in I \cap \mathscr{D}(f)$.
Proof.   There exists a neighborhood I of a such that

$$|f(x) - A| < \frac{A}{2} \quad \left(|f(x) - A| < -\frac{A}{2} \right) \quad \text{whenever } a \neq x \in I \cap \mathscr{D}(f).$$

By a routine calculation on these inequalities the desired conclusion follows.

7.25   Theorem.   If $\lim\limits_{a} f \in R$, then $\lim\limits_{a} |f| = |\lim\limits_{a} f|$. If in addition $k \in R$,
then $\lim\limits_{a} k f = k \lim\limits_{a} f$.
Proof.   We let $\lim\limits_{a} f = A$, invoke Lemma 7.23 to find a neighborhood
I of a such that $f(x) \in R$ whenever $a \neq x \in I \cap \mathscr{D}(f)$, and consider an arbitrary sequence X with $\lim\limits_{n} X_n = a$, $\mathscr{R}(X) \subset (I \cap \mathscr{D}(f) - \langle a \rangle)$. We see by Lemma 7.4 that $\lim\limits_{n} f(X_n) = A$, apply Theorem 5.16 to obtain $\lim\limits_{n} |f(X_n)| = |A|$ and $\lim\limits_{n} k f(X_n) = k \lim\limits_{n} f(X_n)$, and finally apply Lemma 7.13 to complete the proof.

7.26   Theorem.   If f and g are functions with a common domain S such that $\lim\limits_{a} f \in R$ and $\lim\limits_{a} g \in R$, then

(i) $\lim\limits_{a} (f + g) = \lim\limits_{a} f + \lim\limits_{a} g$

(ii) $\lim_a (f - g) = \lim_a f - \lim_a g$

(iii) $\lim_a f \cdot g = (\lim_a f) \cdot (\lim_a g)$

(iv) $\lim_a f/g = (\lim_a f)/(\lim_a g)$ if $\lim_a g \neq 0$,

provided the respective functions $f + g$, $f - g$, fg, f/g have S as their domain.

Proof. We consider any neighborhood U of a and an arbitrary sequence X with $\lim_n X_n = a$, $\mathscr{R}(X) \subset (U \cap S - \langle a \rangle)$. In accordance with Lemma 7.4 and Theorem 5.16, we see that

$$\lim_n (f(X_n) + g(X_n)) = \lim_n f(X_n) + \lim_n g(X_n) = \lim_a f + \lim_a g$$

$$\lim_n (f(X_n) - g(X_n)) = \lim_n f(X_n) - \lim_n g(X_n) = \lim_a f - \lim_a g$$

$$\lim_n f(X_n) \cdot g(X_n) = \left(\lim_n f(X_n)\right) \cdot \left(\lim_n g(X_n)\right) = (\lim_a f) \cdot (\lim_a g)$$

$$\lim_n f(X_n)/g(X_n) = \left(\lim_n f(X_n)\right)/\left(\lim_n g(X_n)\right) = (\lim_a f)/(\lim_a g),$$

respectively. From the arbitrary nature of the sequence X and Lemma 7.13, the truth of (i), (ii), (iii), and (iv) becomes apparent.

7.27 Theorem. If f and g are functions with a common domain $S \subset Re$, such that

(i) $\lim_a f = A \in R$ and $\lim_a g = +\infty(-\infty)$, then $\lim_a (f + g) = \lim_a f + \lim_a g = +\infty(-\infty)$; if in particular $A \neq 0$, then $\lim_a fg = (\lim_a f)(\lim_a g)$.

(ii) $\lim_a f = +\infty$ or $-\infty$ and $\lim_a g = +\infty$ or $-\infty$, then $\lim_a f \cdot g = (\lim_a f) \cdot (\lim_a g)$. If in particular $\lim_a f = \lim_a g = +\infty$ $(-\infty)$, then $\lim_a (f + g) = \lim_a f + \lim_a g = +\infty$ $(-\infty)$.

Proof (i). Take a neighborhood U of a and let X be any sequence with $\lim_n X_n = a$ and $\mathscr{R}(X) \subset (U \cap S - \langle a \rangle)$; then $\lim_n f(X_n) = A$ and $\lim_n g(X_n) = +\infty$ $(-\infty)$ by Lemma 7.4. Theorems 5.17 and 5.20 allow us to infer that $\lim_n (f(X_n) + g(X_n)) = \lim_n f(X_n) + \lim_n g(X_n) = \lim_a f + \lim_a g = +\infty$ $(-\infty)$ and that

$$\lim_n f(X_n) \cdot g(X_n) = \left(\lim_n f(X_n)\right) \cdot \left(\lim_n g(X_n)\right) = (\lim_a f) \cdot (\lim_a g)$$

under the stronger hypotheses above. Lemma 7.13 may now be applied to complete the task.

(ii). Taking U and X as in (i) above, we see that $\lim_n f(X_n) = +\infty$ or $-\infty$ and $\lim_n g(X_n) = +\infty$ or $-\infty$, by Lemma 7.4, whence Theorems 5.17 and 5.20 yield $\lim_n f(X_n) \cdot g(X_n) = (\lim_a f) \cdot (\lim_a g)$; and with the stronger hypotheses, $\lim_n (f(X_n) + g(X_n)) = \lim_a f + \lim_a g = +\infty\,(-\infty)$. Lemma 7.13 now yields the desired conclusion.

7.28   Theorem. If $\lim_a f = +\infty(-\infty)$, then $\lim_a (1/f) = 0$.

Proof. We assume that ϵ is an arbitrary positive number. We see from Definition 7.1 that there exists a neighborhood I of a such that $0 < 1/\epsilon < |f(x)|$ whenever $a \neq x \in I \cap \mathscr{D}(f)$, and recalling the conventions of § 5.10 we deduce that

$$(1) \qquad\qquad 0 \leq \left| \frac{1}{f(x)} \right| < \epsilon$$

for each such x. Now $\mathscr{D}(1/f) \subset \mathscr{D}(f)$ because of our conventions of § 7.19, thus (1) holds for all x, $a \neq x \in I \cap \mathscr{D}(1/f)$. Also from (1) we see that $I \cap \mathscr{D}(f) \subset (\mathscr{D}(1/f) \cup \langle a \rangle)$. Using Corollaries 7.12 and 3.12 we infer that a is a cluster point of $\mathscr{D}(1/f) \cup \langle a \rangle$, and thus of $\mathscr{D}(1/f)$ itself. From (1) we now conclude that $\lim_a (1/f) = 0$.

7.29   Theorem. If f is such a function that $0 < f(x) < +\infty(-\infty < f(x) < 0)$ whenever $x \in \mathscr{D}(f)$ and $\lim_a (1/f) = 0$, then $\lim_a f = +\infty\,(-\infty)$.

Proof. From our hypotheses we see that $0 < (1/f(x)) \in R\,(0 > (1/f(x)) \in R)$ whenever $x \in \mathscr{D}(f)$. Also a is a cluster point of $\mathscr{D}(1/f)$ in view of our conventions in §7.14 and §7.19. We suppose M is an arbitrary finite positive number. According to Definition 7.1, there exists a neighborhood I of a such that $|1/f(x)| < 1/M$ whenever $a \neq x \in I \cap \mathscr{D}(1/f) = I \cap \mathscr{D}(f)$. In the light of our hypotheses, we see that this means $f(x) > M\,(f(x) < -M)$ for each such x, whence $\lim_a f = +\infty\,(-\infty)$.

7.30   Some Applications. The results established in Theorems 7.20 to 7.29 inclusive all required that the functions concerned have a common domain, yet it is frequently necessary and possible to use these results on functions not possessing a common domain. It would be too involved and lengthy to prove these theorems under more general hypotheses, but we shall discuss here some situations wherein the conclusions are still valid, and they point out how one could make appropriate modifications in our theorems to take care of still other situations.

We start by considering functions f and g defined on a common domain $S \subset Re$, with a cluster point $a \in S$. We suppose that $f(x)$ and $g(x)$ are finite real numbers for each $x \in (S - \langle a \rangle)$ and that $f(a) = +\infty$, $g(a) = -\infty$. From our conventions in § 7.19 it is clear that $\mathscr{D}(f + g) = S - \langle a \rangle \neq S$. We suppose that $\lim_{a} f \in R$ and $\lim_{a} g \in R$. We would like to infer that $\lim_{a} (f + g) = \lim_{a} f + \lim_{a} g$, but we cannot apply Theorem 7.26 directly. However, if we let $T = S - \langle a \rangle$, and observe that $f + g = ((f + g) \mid T) = (f \mid T) + (g \mid T)$, $\lim_{a} (f \mid T) = \lim_{a} f$, $\lim_{a} (g \mid T) = \lim_{a} g$, then it is possible to use Theorem 7.26 to conclude that $\lim_{a} (f + g) = \lim_{a} f + \lim_{a} g$. A variation of the situation just considered is the following: f is defined on S, g on $S - \langle a \rangle$, $f(x) \in R$ whenever $x \in S$, $g(x) \in R$ whenever $x \in (S - \langle a \rangle)$, $\lim_{a} f \in R$, $\lim_{a} g \in R$. Clearly $f + g$ is defined on $S - \langle a \rangle$ and takes real values there. By considerations of the type just made, it is again possible to infer that $\lim_{a} (f + g) = \lim_{a} f + \lim_{a} g$. It is clear that these methods could be applied to other parts of Theorem 7.26 and to other theorems from 7.20 to 7.29 inclusive. Due to Corollary 7.16 and these last observations, it follows that if there exists a neighborhood I of a in which the functions $(f \mid I - \langle a \rangle)$ and $(g \mid I - \langle a \rangle)$ satisfy the hypotheses of Theorems 7.20 to 7.29, then the conclusions of these theorems are valid for the functions f and g.

7.31   Theorem. If f and g are such functions that $\mathscr{R}(f) \subset \mathscr{D}(g)$, $\lim_{u \to b} g(u) = A \in Re$, $\lim_{x \to a} f(x) = b \in Re$, and there exists a neighborhood I' of a such that $f(x) \neq b$ whenever $a \neq x \in I' \cap \mathscr{D}(f)$, then $\lim_{x \to a} g(f(x)) = b$.

Proof. We let $h = g \circ f$. By virtue of Definition 1.47, $\mathscr{D}(h) = \mathscr{D}(f)$, $\mathscr{R}(h) \subset \mathscr{R}(g)$.

We suppose that J is an arbitrary neighborhood of A; then there is a neighborhood K of b such that

$$(1) \qquad\qquad g(u) \in J$$

whenever $b \neq u \in K$. Also there is a neighborhood I'' of a such that

$$(2) \qquad\qquad f(x) \in K$$

whenever $a \neq x \in I''$. We let $I = I' \cap I''$; I is a neighborhood of a by Lemma 7.11; by hypothesis, $b \neq f(x)$ whenever $a \neq x \in I'$. Thus, putting (1) and (2) together, we infer that $g(f(x)) \in J$ whenever $a \neq x \in I \cap \mathscr{D}(f)$; consequently $\lim_{x \to a} h(x) = \lim_{x \to a} g(f(x)) = A$.

The theorem just proved is useful in many situations. In particular, it allows us to equate limits of functions at $+\infty$ $(-\infty)$ to limits of certain associated functions at 0. Since it is often simpler to deal with finite numbers than with infinite ones, the preceding theorem enables us to evaluate certain limits conveniently. We will see how this comes about in Theorem 7.35, after some preliminary lemmas have been proved, first recalling § 7.17.

7.32 Lemma. If $+\infty(-\infty)$ is a cluster point of the set $S \subset Re$, then $+\infty$ $(-\infty)$ is a cluster point of S_0^+ (S_0^-), and conversely.
Proof. We suppose that K is an arbitrary finite positive number. If $K < x \in S$ $(-K > x \in S)$, then clearly $K < x \in S_0^+$ $(-K > x \in S_0^-)$ and conversely. Consequently

$$S \cap \{x \mid x > K\} = S_0^+ \cap \{x \mid x > K\}$$
$$(S \cap \{x \mid x < -K\} = S_0^- \cap \{x \mid x < -K\}).$$

From this we see that if the set on the left of the equality sign is infinite, so is that on the right and conversely, and the lemma is proved.

7.33 Definition. If $S \subset R$, we agree to let $\bar{S} = \{x \mid \text{there exists } y \in S$ such that $y \neq 0$ and $x = 1/y\}$.
In brief, $\bar{S}$ is the set of reciprocals of members of S, except 0, in case $0 \in S$.

7.34 Lemma. If $S \subset R$ and $+\infty$ $(-\infty)$ is a cluster point of S, then 0 is a cluster point of $\bar{S}_0^+(\bar{S}_0^-)$ and conversely.
Proof. We assume $+\infty(-\infty)$ is a cluster point of S; by Lemma 7.32, $+\infty$ $(-\infty)$ is a cluster point of S_0^+ (S_0^-). If $\epsilon > 0$, there exists $t \in S_0^+$ $(t \in S_0^-)$ such that $0 < 1/\epsilon < t < +\infty$ $(-\infty < t < -1/\epsilon < 0)$. Let $x = 1/t$; clearly $x \in \bar{S}_0^+$ $(x \in \bar{S}_0^-)$ and $0 < x = 1/t < \epsilon$ $(-\epsilon < x = 1/t < 0)$; by Theorem 3.23, 0 is a cluster point of $\bar{S}_0^+$ $(\bar{S}_0^-)$.
If 0 is a cluster point of $\bar{S}_0^+$ $(\bar{S}_0^-)$ and $0 < M < +\infty$, there exists $x \in \bar{S}_0^+$ $(x \in \bar{S}_0^-)$ such that $0 < x < 1/M (-1/M < x < 0)$ and there exists $t \in S_0^+$ $(t \in S_0^-)$ such that $x = 1/t$. Then $0 < M < t < +\infty (-\infty < t < -M < 0)$; by Theorem 5.8, $+\infty$ $(-\infty)$ is a cluster point of S.

7.35 Theorem. If g is a function whose domain S and range are subsets of Re, $+\infty$ $(-\infty)$ is a cluster point of S, h is defined on $\bar{S}$ so that $h(t) = g(1/t)$ whenever $t \in \bar{S}$ and $\lim_{t \to 0+} h(t) = A \in Re$ $(\lim_{t \to 0-} h(t) = A)$, then $\lim_{x \to +\infty} g(x) = A$ $(\lim_{x \to -\infty} g(x) = A)$.
Proof. We shall prove only that if $\lim_{t \to 0+} h(t) = A$, then $\lim_{x \to +\infty} g(x) = A$; the proof that $\lim_{t \to 0-} h(t) = A$ implies $\lim_{x \to -\infty} g(x) = A$ can be achieved by replacing S_0^+ and $\bar{S}_0^+$ everywhere they occur in the proof below by S_0^- and $\bar{S}_0^-$, respectively.

We define f on S_0^+ so that $f(x) = 1/x$ whenever $x \in S_0^+$. Since $\mathcal{R}(f) = \overline{S}_0^+ \subset \overline{S}$ we may form the function $(h \mid \overline{S}_0^+) \circ f$ in accordance with Definition 1.47. For any $x \in S_0^+$ we have

$$((h \mid \overline{S}_0^+) \circ f)(x) = h(f(x)) = h(1/x) = g(x) = (g \mid S_0^+)(x).$$

Hence $(g \mid S_0^+) = (h \mid \overline{S}_0^+) \circ f$ due to Corollary 1.40. By hypothesis, $\lim_{t \to 0} h(t) = \lim_{t \to 0} (h \mid \overline{S}_0^+)(t) = A$; also $\lim_{x \to +\infty} f(x) = 0$ and $f(x) \neq 0$ for each $x \in S_0^+$. Thus by Theorem 7.31, $\lim_{x \to +\infty} (g \mid S_0^+)(x) = A$. Since S_0^+ is a neighborhood of $+\infty$, then Corollary 7.16 yields $\lim_{x \to +\infty} g(x) = A$.

We show how this last theorem may be applied by considering an example. Suppose that we wish to evaluate $\lim_{x \to -\infty} x \sin(1/x)$. We let $g(x) = x \sin(1/x)$ and replace x everywhere it appears in $g(x)$ by $1/t$, thus obtaining the new function h satisfying $h(t) = (1/t) \sin t = (\sin t)/t$ for each real $t \neq 0$. According to our theorem, if we can evaluate $\lim_{t \to 0-} h(t)$, then the value so obtained will be the value of $\lim_{x \to -\infty} g(x)$. In this particular case $\lim_{t \to 0-} h(t) = 1$, and our problem is solved.

In some applications of the theorem, it may turn out to be difficult or impossible to evaluate $\lim_{t \to 0+} h(t) \left(\lim_{t \to 0-} h(t) \right)$. In such cases we have to look for other methods of evaluating the given limit expression. Although our theorem as we have proved it does not allow us to infer this, it is also a fact that if $\lim_{t \to 0+} h(t) \left(\lim_{t \to 0-} h(t) \right)$ can be shown not to exist, then the same is true of $\lim_{t \to +\infty} g(x) \left(\lim_{t \to -\infty} g(x) \right)$ (cf. Problem 69 in the Exercises at the end of the chapter).

As was mentioned earlier, there is for finite limits of functions of a real variable an analogue, in fact a generalization, of the Cauchy criterion for convergence of sequences; it even goes by the name of Cauchy criterion. We shall prove this now.

7.36   Theorem.   If $a \in Re$, f is a function with $\mathcal{D}(f) \subset Re$, $\mathcal{R}(f) \subset Re$, a is a cluster point of $\mathcal{D}(f)$, and for an arbitrary positive number ϵ there exists a neighborhood I of a such that $|f(x') - f(x'')| < \epsilon$ whenever $a \neq x' \in I \cap \mathcal{D}(f)$ and $a \neq x'' \in I \cap \mathcal{D}(f)$, then $\lim_a f$ exists and is a finite real number, and conversely.

Proof.   We shall establish the converse first. We suppose that $\lim_a f = A \in R$. Then if $\epsilon > 0$, there exists a neighborhood I of a such that

$$(1) \qquad\qquad |f(x) - A| < \frac{\epsilon}{2}$$

whenever $a \neq x \in I \cap \mathscr{D}(f)$. Now if $a \neq x' \in I \cap \mathscr{D}(f)$ and $a \neq x'' \in I \cap \mathscr{D}(f)$, we see from (1) that

$$|f(x') - f(x'')| \leq |f(x') - A| + |A - f(x'')| < \frac{\epsilon}{2} + \frac{\epsilon}{2} = \epsilon,$$

as we wished to show.

We now turn to the proof that the stated criterion guarantees that $\lim_a f$ exists in R. We see from Lemma 7.3 that there exists a sequence X with $\lim_n X_n = a$ and $\mathscr{R}(X) \subset (\mathscr{D}(f) - \langle a \rangle)$. We suppose that ϵ is an arbitrary positive number and use our hypotheses to ascertain the existence of a neighborhood I of a such that

$$(2) \qquad\qquad |f(x') - f(x'')| < \frac{\epsilon}{2}$$

whenever $a \neq x' \in I \cap \mathscr{D}(f)$ and $a \neq x'' \in I \cap \mathscr{D}(f)$.

Now from our choice of X and the definition of sequential limits, there exists a positive integer N such that $X_p \in I$ whenever $N < p \in P$. Thus if $N < m, n \in P$, we see from (2) that

$$(3) \qquad\qquad |f(X_m) - f(X_n)| < \frac{\epsilon}{2} < \epsilon.$$

This is just the condition that $\{f(X_n)\}$ be a Cauchy sequence. Thus $\{f(X_n)\}$ converges; we let $A = \lim_n f(X_n) \in R$. We shall use the fact that this strategically located sequence converges to A to show that $\lim_a f = A$. Since $\epsilon > 0$, there exists a positive integer M such that

$$(4) \qquad\qquad |f(X_n) - A| < \frac{\epsilon}{2}$$

whenever $M < n \in P$. Now we consider an arbitrary number x such that $a \neq x \in I \cap \mathscr{D}(f)$. From (2) and (4) we see that

$$|f(x) - A| \leq |f(x) - f(X_{M+N})| + |f(X_{M+N}) - A| < \frac{\epsilon}{2} + \frac{\epsilon}{2} = \epsilon.$$

Thus $\lim_a f = A$, as was to be shown.

We turn now to one of the most important and useful concepts in analysis, namely, that of continuity.

7.37 Definition. If f is a function whose domain and range are subsets of Re, and $a \in R \cap \mathscr{D}(f)$, then we say that f is *continuous* at a if and only if for an arbitrary positive number ϵ there exists $\delta > 0$ such that $|f(x) - f(a)| < \epsilon$ whenever $|x - a| < \delta$ and $x \in \mathscr{D}(f)$; equivalently, for each neighborhood J of $f(a)$ there exists a neighborhood I of a such that $f(x) \in J$ whenever $x \in I \cap \mathscr{D}(f)$. If $T \subset \mathscr{D}(f)$, and if f is continuous at each point $x \in T$, then we say that f is *continuous on* T.

Although there is a considerable similarity between this and Definition 7.1, there are some noteworthy differences. First, the above definition demands that a be a finite real number; it does not require that a be a cluster point of $\mathscr{D}(f)$ but instead insists that a belong to $\mathscr{D}(f)$. Finally, whereas Definition 7.1 specifically excludes reference to the value (if any) of f at a, the above specifically makes demands on the value of f at a; $f(a)$ must, in particular, clearly be finite.

7.38   Definition.   If $a \in S \subset Re$, we say that a is an *isolated* point of S if and only if a is not a cluster point of S.

7.39   Theorem.   If $a \in S \subset Re$, then a is an isolated point of S if and only if there exists a neighborhood I of a such that $I \cap S = \langle a \rangle$.
Proof.   This is an immediate consequence of Theorems 3.23 and 5.8.

7.40   Theorem.   If f is a function whose domain and range are subsets of Re, a is a finite isolated point of $\mathscr{D}(f)$, and $f(a)$ is finite, then f is continuous at a.
Proof.   By Theorem 7.39 there is a neighborhood I of a such that $I \cap S = \langle a \rangle$. Thus if x is an arbitrary member of $I \cap \mathscr{D}(f)$, it follows that $x = a$, whence if $\epsilon > 0$, then $|f(x) - f(a)| = 0 < \epsilon$, and so f is continuous at a.

7.41   Theorem.   If f is a function whose domain and range are subsets of Re, $a \in \mathscr{D}(f)$ and is a finite cluster point of $\mathscr{D}(f)$, and $f(a)$ is finite, then f is continuous at a if and only if $\lim_{x \to a} f(x) = f(a)$.
Proof.   If f is continuous at a, and ϵ is an arbitrary positive number, then there exists $\delta > 0$ such that $|f(x) - f(a)| < \epsilon$ whenever $|x - a| < \delta$ and $x \in \mathscr{D}(f)$, from which it is apparent that $\lim_{x \to a} f(x) = f(a)$.

On the other hand, if $f(a)$ is finite, $\lim_{x \to a} f(x) = f(a)$, and ϵ is an arbitrary positive number, then there exists $\delta > 0$ such that $|f(x) - f(a)| < \epsilon$ whenever $0 < |x - a| < \delta$ and $x \in \mathscr{D}(f)$. However, if $x = a$, then we clearly have $|f(x) - f(a)| = 0 < \epsilon$, therefore f is continuous at a.

7.42   Theorem.   If f is a function continuous at a and if $a \in T \subset \mathscr{D}(f)$, then $(f \mid T)$ is continuous at a.
Proof.   If a is an isolated point of T, then since $(f \mid T)(a) = f(a)$ is finite due to the continuity of f at a, the continuity of $(f \mid T)$ at a is assured by Theorem 7.40.

If a is a cluster point of T, then it is clearly a cluster point of $\mathscr{D}(f)$, so by Theorem 7.41 $\lim_{x \to a} f(x) = f(a)$. Thus by Lemma 7.15, $\lim_{x \to a} (f \mid T)(x) = f(a) = (f \mid T)(a)$, from which the desired continuity of $(f \mid T)$ follows by a second application of Theorem 7.41.

7.43   Theorem.   If f is a function continuous at a finite point $a \in \mathscr{D}(f)$ and $k \in R$, then $|f|$ and kf are continuous at a.

Proof.   If a is an isolated point of $\mathscr{D}(f)$, then since $|f|$ and kf have finite values there, the desired conclusion follows from Theorem 7.40.

If a is a cluster point of $\mathscr{D}(f)$, then by Theorem 7.41, $\lim_{x \to a} f(x) = f(a)$, and so $|f|$ and kf are continuous at a by virtue of Theorem 7.25 and an additional application of Theorem 7.41.

7.44   Theorem.   If f and g are functions with a common domain $S \subset Re$, each continuous at $a \in S$, then $f + g, f - g$, and $f \cdot g$ are continuous at a; so also is f/g if $g(a) \neq 0$.

Proof.   If a is an isolated point of S, the theorem follows from Theorem 7.40. If a is a cluster point of S, we use Theorem 7.41 to infer that $\lim_{x \to a} f(x) = f(a)$ and $\lim_{x \to a} g(x) = g(a)$. The desired conclusion now follows from Theorem 7.26 and another application of Theorem 7.41.

7.45   Theorem.   If f is a function whose domain and range are subsets of Re, a is a finite point of $\mathscr{D}(f)$, I is a neighborhood of a, and $f(a)$ is a finite real number, then f is continuous at a if and only if $\lim_{n} f(X_n) = f(a)$ whenever X is a sequence whose range is contained in $I \cap \mathscr{D}(f)$ and $\lim_{n} X_n = a$.

Proof.   We suppose first that f is continuous at a, and we take an arbitrary sequence X such that $\mathscr{R}(X) \subset I \cap \mathscr{D}(f)$ and $\lim_{n} X_n = a$. If ϵ is an arbitrary positive number, then there exists $\delta > 0$ such that $|f(x) - f(a)| < \epsilon$ whenever $|x - a| < \delta$. Also, there exists a positive integer N such that $|X_n - a| < \delta$ whenever $N < n \in P$. Hence $|f(X_n) - f(a)| < \epsilon$ for each such n; consequently $\lim_{n} f(X_n) = f(a)$ as required.

Next, we suppose that $\lim_{n} f(X_n) = f(a)$ whenever X is such a sequence that $\mathscr{R}(X) \subset I \cap \mathscr{D}(f)$ and $\lim_{n} X_n = a$. In case a is an isolated point of $\mathscr{D}(f)$, we know from Theorem 7.40 that f is continuous at a. If a is a cluster point of $\mathscr{D}(f)$, and if X is an arbitrary sequence such that $\mathscr{R}(X) \subset I \cap \mathscr{D}((f) - \langle a \rangle)$ and $\lim_{n} X_n = a$, then the assumption of this paragraph assures us that $\lim_{n} f(X_n) = f(a)$, whence by Lemma 7.13, $\lim_{x \to a} f(x) = f(a)$, and so by Theorem 7.41, f is continuous at a.

7.46   Theorem.   If $h = g \circ f$, where g is a function continuous at b, and f is such a function that $\lim_{x \to a} f(x) = b$, then $\lim_{x \to a} h(x) = \lim_{x \to a} g(f(x)) = g(b)$.

Proof. We remark first that this is not a simple corollary of Theorem 7.31. For, although we demand more of g here, we also demand a little less of f; namely, $f(x)$ may coincide with b without any restriction on x.

We suppose that ϵ is an arbitrary positive number. Then, due to the continuity of g at b, there exists $\eta > 0$ such that

(1) $$|g(u) - g(b)| < \epsilon$$

whenever $|u - b| < \eta$ and $u \in \mathscr{D}(g)$. Also, there exists a neighborhood I of a such that, whenever $a \neq x \in I \cap \mathscr{D}(f)$,

(2) $$|f(x) - b| < \eta$$

Combining (1) and (2) we see that $|h(x) - g(b)| = |g(f(x)) - g(b)| < \epsilon$ whenever $a \neq x \in I \cap \mathscr{D}(f)$; thus $\lim_{x \to a} h(x) = \lim_{x \to a} g(f(x)) = g(b)$.

7.47 Corollary. If h and g satisfy the hypotheses of Theorem 7.46 and f is continuous at a, then h is continuous at a.

Proof. In this special case we see that a must be a finite real number and $b = f(a) \in R$. By Theorem 7.40, we need consider only the case that a is a cluster point of $\mathscr{D}(f)$; applying Theorem 7.46 we conclude that $\lim_{x \to a} h(x) = g(b) = g(f(a)) = h(a)$, so that h is continuous by Theorem 7.41.

Many of the most interesting and useful results in mathematics concern functions that are continuous on particular kinds of sets, especially those whose domains are *closed bounded intervals*, shortly to be defined.

7.48 Definition. A subset I of Re is called an *interval* if and only if there exist members a and b of Re such that $a < b$ and

$$\{x \mid a < x < b\} \subset I \subset \{x \mid a \leq x \leq b\}.$$

I is said to be *open* if and only if $I = \{x \mid a < x < b\}$;
I is said to be *closed* if and only if $I = \{x \mid a \leq x \leq b\}$;
I is said to be *half-open to the left* (*right*) if and only if $I = \{x \mid a < x \leq b\}$ ($I = \{x \mid a \leq x < b\}$).
We shall refer to a and b as *left* and *right endpoints of I*, respectively, and we shall prove their uniqueness presently.

7.49 Lemma. If I is an interval, then its left and right endpoints are unique.

Proof. We assume that a, a', b and b' all belong to Re, $a < b$, $a' < b'$, and

(1)
$$\{x \mid a < x < b\} \subset I \subset \{x \mid a \leq x \leq b\};$$
$$\{x \mid a' < x < b'\} \subset I \subset \{x \mid a' \leq x \leq b'\}.$$

We suppose, if possible, that $a' < a$ ($b < b'$). Then there exists a real number r such that $a' < r < a$ and $a' < r < b'$ ($b < r < b'$ and $a' < r < b'$). From the first relation in (1) we see that $r \notin \{x \mid a \leq x \leq b\}$; hence $r \notin I$. From the second relation in (1), we see that $r \in \{x \mid a' < x < b'\}$; hence $r \in I$. This contradiction rules out the possibility that $a' < a$ ($b < b'$). Similarly, it follows that we cannot have $a < a'$ ($b' < b$). Thus $a = a'$ and $b = b'$, and the proof is complete.

This result justifies the following definition.

7.50 Definition. In the terminology of Definition 7.48, we shall say that *a is the left endpoint of I* and *b is the right endpoint of I*. We shall say that x is an *interior point of I* if and only if $a < x < b$.

If both a and b are finite, we shall say that I is a *finite* or *bounded interval of length* $b - a$; we shall say that $(a + b)/2$ is the *midpoint* or *center* of I. If a or b is infinite, we shall say that I is an *infinite* or *unbounded interval of infinite length.*

7.51 Corollary. If I is an interval, then inf I and sup I are its left and right endpoints, respectively.

Proof. By Definition 7.48, there exist members a and b of Re such that $a < b$ and $\{x \mid a < x < b\} \subset I \subset \{x \mid a \leq x \leq b\}$.

Clearly, for any $x \in I$ we must have $a \leq x \leq b$; hence a is a lower bound and b an upper bound for I. On the other hand, if $a < u$ ($u < b$), then there exists $x \in Re$ such that $a < x < u$ and $a < x < b$, ($u < x < b$ and $a < x < b$), whence $x \in I$ and so u is not a lower (upper) bound for I. Therefore $a = \inf I$ and $b = \sup I$ as stated above.

7.52 Lemma. If I is a closed interval and X is such a sequence that $X_n \in I$ whenever $n \in P$, then there exists a subsequence Y of X such that $\lim_n Y_n$ exists and belongs to I.

Proof. By Theorems 6.6 and 6.7, there exists a subsequence Y of X such that $\lim_n Y_n = \limsup_n X_n$; and by Definition 4.27, there exists a strictly increasing positive integer valued sequence f such that $Y = X \circ f$. Hence $Y_n = X(f(n))$ for each $n \in P$; and if a and b are the left and right endpoints, respectively, of I, then $a \leq Y_n \leq b$ for each such n. Consequently, $a \leq \lim_n Y_n \leq b$, and so $\lim_n Y_n \in I$.

7.53 Theorem. If f is a function whose domain is a bounded closed interval I, and if f is continuous on I, then f is bounded. (Recall Definition 4.6.)

Proof. We shall assume the contrary and show that a contradiction results. Then, given any finite number $M \geq 0$, there exists $x \in I$ such that

$|f(x)| \geq M$. Hence, if we so define the sequence of sets T that $T_n = \{x \mid x \in I \text{ and } |f(x)| \geq n\}$ whenever $n \in P$, it follows that $T_n \neq \varnothing$ for each such n; and so using the form of the axiom of choice discussed at the end of Chapter 1, there exists a sequence X such that $X_n \in T_n$ for each $n \in P$; i.e. $X_n \in I$ and $|f(X_n)| \geq n$ for each such n.

According to Lemma 7.52 there exists a subsequence Y of X such that $\lim_n Y_n \in I$. We let $c = \lim_n Y_n$, and we let g be such a strictly increasing positive integer valued sequence that $Y = X \circ g$. Recalling Lemma 4.28, we infer that $|f(Y_n)| = |f(X(g(n)))| \geq g(n) \geq n$ for each $n \in P$. From Theorem 7.45 and this last result, we conclude that $+\infty = \lim_n f(Y_n) = f(c)$, an impossibility since f is continuous on I and $c \in I$. $_n$

The preceding theorem need not be true if the interval I is not closed. To show this, we consider the function f so defined on $I = \{x \mid 0 < x \leq 1\}$ that $f(x) = 1/x$ whenever $x \in I$. Clearly f is continuous on I, but it is not bounded on I. In this case it can be seen that the sequence X of the theorem converges to 0, and since in this case $0 \notin I$, the proof breaks down.

7.54 Theorem. If I is a bounded closed interval, f is a function continuous on I, and $M = \sup \mathcal{R}(f)$ $(m = \inf \mathcal{R}(f))$, then there exists u such that $u \in I$ and $f(u) = M$ $(f(u) = m)$.[6]
Proof. By Theorem 7.53, M (m) is a finite real number. We define the sequence T so that for each $n \in P$,

$$T_n = \left\{x \mid x \in I, M - \frac{1}{n} < f(x) \leq M\right\}$$

$$\left(T_n = \left\{\{x \mid x \in I, m \leq f(x) < m + \frac{1}{n}\right\}\right).$$

Because of Theorem 2.13, $T_n \neq \varnothing$ for each such n; consequently, by the axiom of choice (recall § 1.51) there exists a sequence X such that for each $n \in P$,

(1) $X_n \in I, M - \frac{1}{n} < f(X_n) \leq M$ $\left(m \leq f(X_n) < m + \frac{1}{m}\right)$.

From (1) we conclude that $\lim_n f(X_n) = M$ $(\lim_n f(X_n) = m)$.

If a and b are the endpoints of I, then $a \leq X_n \leq b$ for each $n \in P$, and by Lemma 7.52 there exists a subsequence Y of X such that Y converges to a point $c \in I$. Also, $\{f(Y_n)\}$ is a subsequence of $\{f(X_n)\}$ (cf. Definition 4.27); hence by Theorems 4.29 and 7.45, $f(c) = \lim_n f(Y_n) = \lim_n f(X_n) = M$ (m).

[6] Many such values of u may exist.

7.55 Lemma. If f is a function continuous on its domain, which is a closed bounded interval I whose left endpoint is a and whose right endpoint is b, $f(a) > 0$ and $f(b) < 0$, then there exists $c \in I$ such that $f(c) = 0$.[7]
Proof. Since f is continuous at b and $f(b) < 0$, we may apply Lemma 7.24 to infer the existence of a positive number $\delta > 0$ such that $f(x) < f(b)/2 < 0$ whenever $x \in I$ and $0 < |x - b| < \delta$; this inequality clearly holds for $x = b$ also. We may assume that $\delta \leq b - a$. Consequently,

(1) $$f(x) < 0 \quad \text{whenever} \quad a \leq b - \delta < x \leq b.$$

We let $T = \{x \mid x \in I \text{ and } f(x) \geq 0\}$. Evidently $a \in T$; therefore $T \neq \varnothing$. Also from (1) it follows that $b - \delta$ is an upper bound for T; therefore T has a least upper bound c; evidently $a \leq c \leq b - \delta < b$.

We want to show that $f(c) = 0$. There exists a sequence Y such that $c < Y_n < b$ for each $n \in P$, and $\lim_n Y_n = c$ (e.g., $Y_n = c + (b - c)/n$). Since $Y_n \notin T$ for each $n \in P$, then $f(Y_n) < 0$ for each such n, and by Corollary 4.12 and Theorem 7.45, $\lim_n f(Y_n) = f(c) \leq 0$. To complete the proof we need only eliminate the possibility that $f(c) < 0$.

Accordingly, we suppose if possible that $f(c) < 0$. Clearly $a \neq c$, since $f(a) > 0$; hence $a < c < b$. Applying Lemma 7.24 again, we infer the existence of $\delta' > 0$ such that

(2) $$f(x) < \frac{f(c)}{2} < 0$$

whenever $0 < |x - c| < \delta'$ and $x \in I$. It is clear that (2) holds if $x = c$. By Theorem 2.13 there exists $u \in T$ such that $c - \delta' < u \leq c$. Thus we must have $f(u) \geq 0$ since $u \in T$, and $f(u) < 0$ because of (2). This contradiction completes the proof.

7.56 Corollary. Under the hypotheses of Lemma 7.55, if $f(a) < 0$ and $f(b) > 0$, there exists $c \in I$ such that $f(c) = 0$.
Proof. We define g so that $g(x) = -f(x)$ whenever $x \in I$. Then g is continuous on I and clearly $g(a) = -f(a) > 0$, $g(b) = -f(b) < 0$. We apply the preceding lemma to see that there exists $c \in I$ such that $g(c) = 0$; then $f(c) = -g(c) = 0$ as required.

We are now ready to prove the well-known intermediate value theorem for continuous functions.

7.57 Theorem. If f is a function whose domain is a closed bounded interval I, $M = \sup \mathscr{R}(f)$, $m = \inf \mathscr{R}(f)$, and $m \leq v \leq M$, then there exists $u \in I$ such that $f(u) = v$.

[7] Many such values of c may exist.

Proof. If $v = m$ or $v = M$, the desired result follows from Theorem 7.54. Thus we need consider only the case $m < v < M$. We so define g on I that $g(x) = f(x) - v$ whenever $x \in I$. Due to Theorem 7.44, g is continuous on I. By virtue of Theorem 7.54 there exist numbers x' and x'' belonging to I such that $f(x') = M$ and $f(x'') = m$; evidently $x' \neq x''$ since $M \neq m$. If $x' < x''$ ($x'' < x'$) we define $I' = \{x \mid x' \leq x \leq x''\}$ ($I' = \{x \mid x'' \leq x \leq x'\}$), and we let $h = (g \mid I')$; h is continuous on I' by Theorem 7.42, and since

$$h(x') = g(x') = f(x') - v = M - v > 0$$
$$h(x'') = g(x'') = f(x'') - v = m - v < 0,$$

we may use Lemma 7.55 (Corollary 7.56) to infer the existence of $u \in I' \subset I$ such that $h(u) = f(u) - v = 0$, as required.

7.58   **Definition.**   If f is a function whose domain and range are subsets of Re, then f is said to be *strictly increasing* (*strictly decreasing*) if and only if $f(x) < f(x')$ ($f(x') < f(x)$) whenever x and x' are members of $\mathscr{D}(f)$ and $x < x'$; f is said to be *non-decreasing* (*non-increasing*) if and only if $f(x) \leq f(x')$ ($f(x') \leq f(x)$) for each such x and x'.

7.59   **Lemma.**   If f is a strictly increasing (decreasing) function then f^{-1} is also a strictly increasing (decreasing) function.
Proof.   We see that if $u \in \mathscr{D}(f)$, $v \in \mathscr{D}(f)$ and $u \neq v$, then $f(u) \neq f(v)$ due to the strictly increasing (decreasing) nature of f, and so by Theorem 1.44, f^{-1} is a function. If $w \in \mathscr{D}(f^{-1})$, $z \in \mathscr{D}(f^{-1})$, and $w < z$, then we let $u = f^{-1}(w)$, $v = f^{-1}(z)$. Note that $w = f(u)$, $z = f(v)$, and deduce from the strictly increasing (decreasing) nature of f that we must have $u < v$ ($v < u$). Thus f^{-1} is strictly increasing (decreasing), as was to be shown.

7.60   **Theorem.**   If I is an interval, f is a function whose domain is I, and f is continuous and strictly increasing (decreasing) on I, then $\mathscr{R}(f)$ is an interval. Moreover, if I is open, so is $\mathscr{R}(f)$; if the left and right endpoints of I are a and b, respectively, and if $a \in I$, then $f(a)$ is the left (right) endpoint of $\mathscr{R}(f)$; if $b \in I$, then $f(b)$ is the right (left) endpoint of $\mathscr{R}(f)$.
Proof.   We shall prove our theorem on the assumption that f is strictly increasing; the other case can be handled quite analogously by making rather obvious minor modifications in the proof given here or by considering the function $-f$, which must be strictly increasing if f is strictly decreasing and then applying to $-f$ the results we obtain shortly; this leads rather easily to the desired conclusion for f itself.

We let $I^0 = \{x \mid a < x < b\}$,[8] $g = (f \mid I^0)$, $c = \inf \mathscr{R}(g)$, and $d = \sup \mathscr{R}(g)$. We take an arbitrary member y of $\mathscr{R}(g)$; there exists $x \in I^0$ such that $y = g(x) = f(x)$. Now clearly $c \leq y \leq d$. However, it is not possible that either $y = c$ or $y = d$; for if one of these statements were true, since there exist (finite) real numbers x' and x'' such that $a < x' < x < x'' < b$, then from the strictly increasing nature of f we should have $g(x') = f(x') < f(x) = g(x) = c$ or $d = g(x) = f(x) < f(x'') = g(x'')$, both of which are impossible due to the definitions of c and d. Thus $c < y < d$, and consequently

(1) $$\mathscr{R}(g) \subset \{y \mid c < y < d\}.$$

If y is an arbitrary number such that $c < y < d$, then by Theorem 5.5 there exist numbers w and z in $\mathscr{R}(g)$ such that $c \leq w < y < z \leq d$; and by (1) we infer that $c < w < y < z < d$. From this we see that w and z are finite real numbers. From Lemma 7.59 and the fact that $\mathscr{R}(g^{-1}) = \mathscr{D}(g) = I^0$, we see that $a < g^{-1}(w) < g^{-1}(z) < b$. Hence if we let $I' = \{x \mid g^{-1}(w) \leq x \leq g^{-1}(z)\}$, it follows that I' is a bounded subinterval of I, and by two applications of Theorem 7.42 we infer that $h = (g \mid I')$ is continuous on I'. By Theorem 7.57 there exists $x \in I'$ such that $h(x) = y$. Since $I' \subset I^0 \subset I$, then $h(x) = g(x) = f(x)$, whence $y \in \mathscr{R}(g)$. Thus $\{y \mid c < y < d\} \subset \mathscr{R}(g)$, and from (1) we now conclude that

(2) $$\mathscr{R}(g) = \{y \mid c < y < d\}.$$

In case I is an open interval, then $I = I^0$ and $f = (f \mid I^0) = g$, and $\mathscr{R}(f) = \mathscr{R}(g)$. Thus we see from (2) that $\mathscr{R}(f)$ is an open interval in this case. It may be bounded or not, depending on the nature of f.

If $a \in I$, then due to the strictly increasing nature of f on I, we have $f(a) < f(x) = g(x)$ for each $x \in I^0$, whence $f(a)$ is a lower bound for $\mathscr{R}(g)$, and thus $f(a) \leq c$. Now there exists a sequence X taking all its values in I^0 such that $\lim_n X_n = a$.[9] Clearly $f(X_n) = g(X_n) \geq c$ for each $n \in P$, and so by Theorem 7.4 and Corollary 5.15, $f(a) = \lim_n f(X_n) \geq c$. Consequently, $f(a) = c$. In entirely analogous fashion it follows that if $b \in I$, then $f(b) = d$.

We now see that if $I = \{x \mid a \leq x < b\}$, then

$$\mathscr{R}(f) = \mathscr{R}(g) \cup \langle f(a) \rangle = \{y \mid f(a) \leq y < d\};$$

if

$$I = \{x \mid a < x \leq b\},$$

then

$$\mathscr{R}(f) = \mathscr{R}(g) \cup \langle f(b) \rangle = \{y \mid c < y \leq f(b)\};$$

---

[8] I^0 is the set of all interior points of I and is often called the *interior* of I. It may be a proper subset of or coincide with I.

[9] For example, $\{a + (b - a)/2n\}$ is such a sequence.

finally, if
$$I = \{x \mid a \le x \le b\},$$
then
$$\mathscr{R}(f) = \mathscr{R}(g) \cup \langle f(a) \rangle \cup \langle f(b) \rangle = \{y \mid f(a) \le y \le f(b)\}.$$

7.61   Theorem.   Under the hypotheses of Theorem 7.60, f^{-1} is continuous on its domain J.

Proof.   We shall prove this theorem on the assumption that f is strictly increasing; the other case can be handled analogously by making small and rather obvious modifications in the proof we shall give.

From Lemma 7.59 we see that f^{-1} is a strictly increasing function, and in the notation of Theorem 7.60, since $\mathscr{R}(f) = J = \mathscr{D}(f^{-1})$, we have as a result of that theorem $\{y \mid c < y < d\} \subset J \subset \{y \mid c \le y \le d\}$.

We consider an arbitrary point y interior to J; $c < y < d$. We let $x = f^{-1}(y)$, so that $y = f(x)$. Recalling the proof of the preceding theorem, we see that x must be an interior point of I; $a < x < b$.

We take an arbitrary finite positive number ϵ. There exists a finite positive number ϵ' less than all three of the numbers $x - a$, $b - x$, and ϵ. Then

$$(1) \qquad a < x - \epsilon' < x = f^{-1}(y) < x + \epsilon' < b.$$

Now we let $s = f(x - \epsilon')$, $t = f(x + \epsilon')$. From (1) and the strictly increasing nature of f we see that $c < s < y < t < d$. We let δ denote the smaller of the two finite positive numbers $y - s$ and $t - y$. We consider an arbitrary number u such that $|y - u| < \delta$. Evidently

$$(2) \qquad c < s \le y - \delta < u < y + \delta \le t < d.$$

From (1) and (2) we infer that

$$(3) \qquad a < f^{-1}(s) = x - \epsilon' < f^{-1}(u) < f^{-1}(t) = x + \epsilon' < b.$$

Combining (1) and (3) we conclude that $|f^{-1}(u) - f^{-1}(y)| < \epsilon' < \epsilon$, whence f^{-1} is continuous at y. Thus f^{-1} is continuous at all interior points of J.

The proof that f^{-1} is continuous at an endpoint of J, in case such a point belongs to J, is achieved by slight modification of the proof just given for continuity at an interior point. We will show how this goes if $a \in I$, in which case $c = f(a) \in J$.

We assume that ϵ is an arbitrary finite positive number. Then there exists a finite positive number ϵ', not exceeding ϵ, such that $a < a + \epsilon' < b$. We let $s = f(a + \epsilon')$, and we let $\delta = s - c > 0$. We consider an arbitrary number u such that

$$(4) \qquad c = f(a) \le u < s = c + \delta < d.$$

From (4) it follows that $f^{-1}(c) = a \leq f^{-1}(u) < f^{-1}(s) = a + \epsilon' < b$. Therefore $|f^{-1}(u) - f^{-1}(c)| < \epsilon' < \epsilon$ for each such u, and so f^{-1} is continuous at c.

By an exactly analogous argument, making small appropriate modifications in this last proof, it can be shown that f^{-1} is continuous at d if $b \in I$. Thus whether J is open, half-open, or closed, f^{-1} is continuous on J.

It is impossible to exaggerate the value and utility of the preceding theorems in mathematical analysis. For instance, consider the function f defined on $I = \{x \mid 0 \leq x < +\infty\}$ so that $f(x) = x^n$ for each $x \in I$, where n is a positive integer. It is readily established that $f(0) = 0$, f is strictly increasing, unbounded above, and continuous on I. Thus it follows from our theorems that $\mathcal{R}(f) = I$, and we can also infer that f^{-1} is a continuous, strictly increasing function unbounded on I. Conventionally, $f^{-1}(y)$ is known as the *nth root of y*, usually written $f^{-1}(y) = y^{1/n}$, and the equations $y = x^n$ and $x = y^{1/n}$ are equivalent whenever $0 \leq x < +\infty$ and $0 \leq y < +\infty$. It may be recalled that in Problem 62 of the Exercises at the end of Chapter 2, we were able to obtain in effect the existence of the inverse function f^{-1}, but nothing more. It is the continuity concept that enabled us to obtain so much more information in the present treatment.

Another important application of these theorems occurs in connection with the exponential function,[10] defined on R and expressed by $f(x) = a^x$ for each $x \in R$, where a is a real number different from 0 and 1. If $a > 1$, it can be shown that f is strictly increasing and continuous on R, and that $\mathcal{R}(f) = I = \{y \mid 0 < y < +\infty\}$. Thus the inverse function f^{-1}, commonly called the *logarithm to the base a*, exists, is continuous and strictly increasing on I, and $\mathcal{R}(f^{-1}) = R$. The equations $y = a^x$ and $x = \log_a y$ are equivalent whenever $x \in R$ and $0 < y < +\infty$.

A similar situation exists with respect to the trigonometric functions. However, as is well known, since these functions are periodic, it is necessary to restrict them to certain definite portions of their domains in order to ensure the existence of the inverse functions. Once this is done, the continuity and other properties of the inverse functions are assured.

Continuity is a pointwise property of a function f. Given a point a in the domain of f at which f is continuous, and given $\epsilon > 0$, there exists $\delta > 0$ such that $|f(x) - f(a)| < \epsilon$ whenever $x \in \mathcal{D}(f)$ and $|x - a| < \delta$. We may speak of δ as a number *associated with a and ϵ*. Now if we consider a different point a' at which f is continuous, keeping ϵ fixed, we know that there exists a number $\delta' > 0$ associated with a' and ϵ, but we should not expect that δ and δ' would necessarily be equal. On the other hand, if

---

[10] We have not defined this function nor developed its properties; we assume familiarity with it, however.

we take any positive number δ'' that does not exceed either δ or δ', then the number δ'' is easily seen to be associated with both points a and a' simultaneously, for the given ϵ. It is not difficult to extend this result; if f is continuous at any finite number of points, then given $\epsilon > 0$, with each such point and the given ϵ there is associated a positive number, and there exists a single number δ that is associated simultaneously with all these points and the given ϵ. The question arises: if a function f is continuous on a set S, is it always possible to find a single positive number δ, for a given $\epsilon > 0$, that is associated with all the points of S? The answer is generally negative. This leads to the following definition and also to a theorem that tells us at least some circumstances in which the answer to our question is affirmative.

7.62 Definition. If f is a function with domain and range contained in R then we say that f is *uniformly continuous* on $\mathscr{D}(f)$ if and only if for each $\epsilon > 0$, there exists $\delta > 0$ such that $|f(x) - f(x')| < \epsilon$ whenever x and x' belong to $\mathscr{D}(f)$ and $|x - x'| < \delta$.

7.63 Theorem. If f is uniformly continuous on $\mathscr{D}(f)$, then f is continuous on $\mathscr{D}(f)$.
Proof. This is an obvious consequence of Definitions 7.62 and 7.37.

7.64 Theorem. If f is a function, $\mathscr{D}(f)$ is a closed bounded interval I and f is continuous on I, then f is uniformly continuous on I.
Proof. We prove this by contradiction. If f were not uniformly continuous on I, there must exist $\epsilon > 0$ such that no matter what positive number δ is taken there exist numbers x' and x'' belonging to I for which $|x' - x''| < \delta$ and $|f(x') - f(x'')| \geq \epsilon$. Thus if we define the sequence T so that $T_n = \{(x', x'') \mid x' \in I, x'' \in I, |x' - x''| < 1/n, |f(x') - f(x'')| \geq \epsilon\}$ for each $n \in P$, then $T_n \neq \varnothing$ for each such n. Hence, using the form of the axiom of choice discussed at the end of Chapter 1, there exists a sequence R of ordered pairs such that $R_n \in T_n$ for each $n \in P$. Corresponding to each such n there exists a unique first term and a unique second term comprising the ordered pair R_n. Thus we may so define the sequences X' and X'' that $(X_n', X_n'') = R_n \in T_n$ for each $n \in P$. We then have

$$(1) \quad \mathscr{R}(X') \subset I, \quad \mathscr{R}(X'') \subset I, \quad |X_n' - X_n''| < \frac{1}{n}, \quad |f(X_n') - f(X_n'')| \geq \epsilon$$

for each $n \in P$.

According to Lemma 7.52 there exists a subsequence Y' of X' such that $\lim_n Y_n' = c \in I$. There exists a strictly increasing sequence g of positive integers such that $Y' = X' \circ g$, in accordance with Definition 4.27. We also define $Y'' = X'' \circ g$. For each positive integer n, we have, from (1) and Lemma 4.28,

(2)
$$|f(Y'_n) - f(Y''_n)| = |f(X'(g(n))) - f(X''(g(n)))| \geq \epsilon$$
$$|Y'_n - Y''_n| = |X'(g(n)) - X''(g(n))| < \frac{1}{g(n)} \leq \frac{1}{n}.$$

From (2) we infer that

$$\lim_n Y''_n = \lim_n (Y''_n - Y'_n) + \lim_n Y'_n = 0 + c = c;$$

so by Theorems 7.45 and 4.8, we conclude that

$$\lim_n |f(Y'_n) - f(Y''_n)| = \lim_n (f(Y'_n) - f(Y''_n)) = f(c) - f(c) = 0.$$

But from (2) and Corollary 4.12, we also see that $\lim_n |f(Y'_n) - f(Y''_n)| \geq \epsilon$;

hence $0 \geq \epsilon$, a contradiction proving the theorem.

7.65   Definition.   If u is a sequence of functions each with domain $S \subset R$ and each taking values in R, then we shall say that the sequence u *converges on* S if and only if for each $x \in S$, $\lim_n u_n(x) \in R$. If U is such a function with domain S that $U(x) = \lim_n u_n(x)$ for each $x \in S$, then we say that the sequence u *converges to* U *on* S, or that U *is the limit of the sequence* u.[11]

According to the definition we have just given, if $x \in S$ and $\epsilon > 0$, then there exists $N \in P$ such that $|u_n(x) - U(x)| < \epsilon$ whenever $N < n \in P$. We may speak of N as a positive integer *associated with x and ϵ*. If we keep ϵ fixed, but take another point $x' \in S$, there will be another positive integer N' associated with x' and ϵ, and in fact the positive integer $N + N'$ will be associated with both x and x' and the given value of ϵ. However, as in the case of continuity compared with uniform continuity, we cannot generally expect that for an arbitrary $\epsilon > 0$ there will exist a single positive integer associated with ϵ and all the points of S simultaneously, when S is an infinite set. In case this does happen, we have a special name for the situation.

7.66   Definition.   If u is a sequence of functions each with domain $S \subset R$ and each taking values in R, and U is a function with domain S and values in R, then we say that the sequence u *converges uniformly to* U *on* S if and only if for each positive number ϵ there exists a positive integer N such that $|u_n(x) - U(x)| < \epsilon$ whenever $N < n \in P$ and $x \in S$; the sequence u

---

[11] The uniqueness of the limit function U is a consequence of the uniqueness of sequential limits.

is said to *converge uniformly on S* if and only if there exists a function V to which the sequence converges uniformly on S.

It is apparent from this definition that if u converges uniformly on S, then it converges on S; and because sequential limits are unique, if we let $U(x) = \lim_n u_n(x)$ for each $x \in S$, then the sequence u converges to U. However, the converse need not be true.[12]

Uniformity conditions of various kinds play a very important role in mathematics. We shall introduce still another concept involving uniformity that bears a close relation to that expressed in the definition just given. To introduce this new idea, we consider a sequence u of functions converging on the set S. Then, for each $x \in S$, $\lim_n u_n(x) \in R$; and so by the Cauchy criterion for convergence of a sequence, given any $\epsilon > 0$ and any $x \in S$, there exists a positive integer N such that $|u_n(x) - u_m(x)| < \epsilon$ whenever $N < m, n \in P$. The same statement is true if we take a different point $x' \in S$; however, we cannot expect that an integer N' associated with x' and ϵ will also be associated with x and ϵ. In case there exists a positive integer N that does not depend on $x \in S$, but only on ϵ, we have a special situation to which we give a special name, as follows.

7.67   Definition.   If u is a sequence of functions with a common domain $S \subset R$, each taking values in R, then we shall say that *u satisfies a uniform Cauchy criterion on S*, or that *u is a uniform Cauchy sequence on S*, if and only if for each $\epsilon > 0$, there exists a positive integer N such that $|u_n(x) - u_m(x)| < \epsilon$ whenever $N < m, n \in P$ and $x \in S$.

We now establish a relationship between uniform convergence on S and uniform Cauchy sequences on S.

7.68   Theorem.   The sequence u of functions converges uniformly on S if and only if it satisfies a uniform Cauchy criterion on S; moreover, in this case the sequence u converges uniformly on S to that function U on S for which $U(x) = \lim_n u_n(x)$ whenever $x \in S$.

Proof.   We suppose first that the sequence u converges uniformly on S. Then there exists a function U on S such that, if ϵ is an arbitrary positive number, there exists a positive integer N for which $|u_p(x) - U(x)| < \epsilon/2$ whenever $N < p \in P$ and $x \in S$. Thus if $x \in S$ and $N < m, n \in P$ we clearly have $|u_n(x) - u_m(x)| \leq |u_n(x) - U(x)| + |U(x) - u_m(x)| < \epsilon$, and the sequence u satisfies a uniform Cauchy criterion on S.

Next, we suppose that the sequence u satisfies a uniform Cauchy criterion on S. This implies that for each $x \in S$, $\{u_n(x)\}$ is a Cauchy sequence, and so converges, which allows us to define a function U on S

---

[12] See the Exercises at the end of this chapter, Problem 91.

in such a way that $U(x) = \lim_n u_n(x)$ for each $x \in S$. We propose to show that the sequence u converges uniformly on S to U.

To this end we let ϵ be an arbitrary positive number. Our hypotheses ensure the existence of a positive integer N such that

$$(1) \qquad\qquad |u_n(x) - u_m(x)| < \frac{\epsilon}{2}$$

whenever $N < m$, $n \in P$ and $x \in S$. We shall show that

$$(2) \qquad\qquad |u_m(x) - U(x)| < \epsilon$$

whenever $N < m \in P$ and $x \in S$; this will establish the uniform convergence of the sequence u to U on S.

Accordingly we consider an arbitrary element $x \in S$ and an arbitrary positive integer m such that $N < m \in P$. Due to the ordinary convergence of the sequence $\{u_n(x)\}$ to $U(x)$ there exists a positive integer N' such that

$$(3) \qquad\qquad |u_n(x) - U(x)| < \frac{\epsilon}{2}$$

whenever $N' < n \in P$. Putting (1) and (3) together, we obtain

$$(4) \quad |u_m(x) - U(x)| \le |u_m(x) - u_{N+N'}(x)| + |u_{N+N'}(x) - U(x)| < \epsilon.$$

This is precisely the relation (2), and the proof is complete.

The technique of the proof just given is worthy of a second look, since variations of it occur frequently in proofs of uniformity. The idea is to obtain a relation of the form (2) involving a positive integer that depends only upon ϵ, not upon the point x appearing therein. We chose N in connection with the relation (1), and it did not depend upon x. However, the integer N' occurring in connection with (3) could not be assumed to depend only upon ϵ; it might also depend upon the point of S under consideration. Thus it might seem that the chain of inequalities in (4) ultimately involves positive integers that depend upon x. However, N' plays a kind of intermediary role in (4) and drops out of the picture entirely since we are concerned only with the first and last terms of (4), which leads to the relation (2) we wanted.

7.69 Theorem. If u is a sequence of continuous functions with a common domain $S \subset R$, converging uniformly on S to U, then U is continuous on S.

Proof. We consider an arbitrary point $x \in S$ and an arbitrary positive number ϵ. There exists a positive integer N such that $|u_n(t) - U(t)| < \epsilon/3$

whenever $N < n \in P$ and $t \in S$. In particular, then

(1)
$$|u_{N+1}(t) - U(t)| < \frac{\epsilon}{3}$$

whenever $t \in S$. Since u_{N+1} is continuous by hypothesis, there exists $\delta > 0$ such that

(2)
$$|u_{N+1}(s) - u_{N+1}(x)| < \frac{\epsilon}{3}$$

whenever $s \in S$ and $|s - x| < \delta$. Thus, putting (1) and (2) together, we see that

$$|U(x) - U(s)| \le |U(x) - u_{N+1}(x)| + |u_{N+1}(x) - u_{N+1}(s)|$$
$$+ |u_{N+1}(s) - U(s)| < \frac{\epsilon}{3} + \frac{\epsilon}{3} + \frac{\epsilon}{3} = \epsilon$$

whenever $s \in S$ and $|x - s| < \delta$. Consequently U is continuous at x, and so is continuous on S.

In the theorem just proved, the value $u_n(x)$ depends on both n and x. We consider the limit expressions $\lim_{n} u_n(x)$ and $\lim_{x \to a} u_n(x)$; in the first, x is held fixed and in the second n is held fixed; and we may now consider the limit expressions obtained from these by allowing x to tend to a in the first and n to tend to infinity in the second. The resulting expressions are, respectively, $\lim_{x \to a} \left(\lim_{n} u_n(x) \right)$ and $\lim_{n} \left(\lim_{x \to a} u_n(x) \right)$. These expressions are called *iterated limits*; they need not be equal, nor even exist. However, under the hypotheses of the preceding theorem, and recalling the notation used, we see that if a belongs to S and is also a cluster point of S, then

$$\lim_{x \to a} \left(\lim_{n} u_n(x) \right) = \lim_{x \to a} U(x) = U(a);$$
$$\lim_{n} \left(\lim_{x \to a} u_n(x) \right) = \lim_{n} \left(u_n(a) \right) = U(a).$$

Thus our theorem may be interpreted as justifying an interchange in the order in which the iterated limits are taken, without altering the value finally obtained. This is frequently a consequence of various forms of uniformity in analysis. If the condition of uniformity is dropped, the iterated limits may not be equal and may not even exist.

We conclude this chapter by establishing an analogue of the concept of limit superior and limit inferior of sequences. We begin with a result that generalizes Theorem 5.24.

7.70 Lemma. If f is a function non-decreasing on $S = \mathcal{D}(f)$ and a is a cluster point of $S_a^-(S_a^+)$, then $\lim_{x \to a-} f(x)$ $\left(\lim_{x \to a+} f(x)\right)$ exists; in fact,

$$\lim_{x \to a-} f(x) = \sup \mathscr{R}(f \mid S_a^-) \; (\lim_{x \to a+} f(x) = \inf \mathscr{R}(f \mid S_a^+))$$

Proof. We let $K = \sup \mathscr{R}(f \mid S_a^-)$ $(k = \inf \mathscr{R}(f \mid S_a^+))$ (recall § 7.17). We suppose that J is an arbitrary neighborhood of $K(k)$. Then by Theorem 5.5 there exists $v \in J \cap \mathscr{R}(f \mid S_a^-)$ $(v \in J \cap \mathscr{R}(f \mid S_a^+))$ and so there exists $u \in S_a^-(u \in S_a^+)$ such that $v = f(u)$. Thus $u < a$ $(a < u)$.

If a is a finite number, we take any finite positive number δ not exceeding $|a - u|$, and we let $I = \{x \mid |x - a| < \delta\}$; if $a = +\infty$ $(a = -\infty)$, then we take any finite positive number M such that $u < M < a = +\infty$. $(-\infty = a < -M < u)$ and let $I = \{x \mid M < x\}$ $(I = \{x \mid x < -M\})$, Clearly, whichever may be the case, if $a \neq x \in I \cap S_a^-$$(a \neq x \in I \cap S_a^+)$, then we have $u < x < a$ $(a < x < u)$, and so, since f is non-decreasing, $f(u) \leq f(x) \leq K$ $(k \leq f(x) \leq f(u))$, whence we see that whether K (k) be finite or not, we must have $f(x) \in J$. Consequently, $\lim_{x \to a-} f(x) = K$ $(\lim_{x \to a+} f(x) = k)$.

7.71 Corollary. Under the assumption of the preceding lemma, except that f is non-increasing instead of non-decreasing,

$$\lim_{x \to a-} f(x) = \inf \mathscr{R}(f \mid S_a^-) \; (\lim_{x \to a+} f(x) = \sup \mathscr{R}(f \mid S_a^+)).$$

Proof. We need only take $g(x) = -f(x)$ whenever $x \in S$ and apply the preceding lemma to infer that

$$\lim_{x \to a-} f(x) = -\lim_{x \to a-} g(x) = -\sup \mathscr{R}(g \mid S_a^-) = \inf \mathscr{R}(f \mid S_a^-)$$

$$\lim_{x \to a+} f(x) = -\lim_{x \to a+} g(x) = -(\inf \mathscr{R}(g \mid S_a^+)) = \sup \mathscr{R}(f \mid S_a^+).[13]$$

7.72 The Limits Superior and Inferior of a Function. We consider a function f with domain $S \subset Re$ taking its values in Re and assume that a is a cluster point of S. For an arbitrary finite positive number d, we define

$$N_d(a) = \{x \mid 0 < |x - a| < d\},$$

or

$$N_d(a) = \left\{x \mid \frac{1}{d} < x < +\infty\right\},$$

or

$$N_d(a) = \left\{x \mid -\infty < x < -\frac{1}{d}\right\},$$

---

[13] Recall Problem 16 in the Exercises at the end of Chapter 5.

respectively, depending on whether $a \in R$, or $a = +\infty$, or $a = -\infty$. For the sake of brevity we shall agree that $T_d = S \cap N_d(a)$ for each such d. We further so define g and G that $g(d) = \inf \mathcal{R}(f \mid T_d)$ and $G(d) = \sup \mathcal{R}(f \mid T_d)$ whenever $0 < d < +\infty$. We note that $T_d \subset T_{d'}$ whenever $0 < d < d' < +\infty$ whence, by Theorem 5.4, g is a non-increasing function and G is a non-decreasing function. Also, since a is a cluster point of S, it follows that $\mathcal{R}(f \mid T_d) \neq \varnothing$ whenever $0 < d < +\infty$, and so by Corollary 5.3, $g(d) \leq G(d)$ for each such d. Applying Lemma 7.70, Corollary 7.71, and Theorem 7.20, we infer that $\lim_{d \to 0+} g(d) \leq \lim_{d \to 0+} G(d)$. We agree to name $\lim_{d \to 0+} g(d)$ the *limit inferior of f at a*, and $\lim_{d \to 0+} G(d)$ the *limit superior of f at a*; we further agree to write these as $\liminf_{x \to a} f(x)$ and $\limsup_{x \to a} f(x)$, respectively.[14]

7.73   Theorem.   If f is a function whose domain and range are subsets of Re, a is a cluster point of $S = \mathscr{D}(f)$ and X is an arbitrary sequence taking values in $S - \langle a \rangle$ for which $\lim_n X_n = a$ and $\lim_n f(X_n) \in Re$, then

$$\liminf_{x \to a} f(x) \leq \lim_n f(X_n) \leq \limsup_{x \to a} f(x).$$

Proof.   Given any finite positive number d, there exists $N \in P$ such that $X_n \in N_d(a)$ whenever $N < n \in P$, since $N_d(a) \cup \langle a \rangle$ is a neighborhood of a, $\lim_n X_n = a$, and X does not take the value a. Thus $\inf \mathcal{R}(f \mid T_d) \leq f(X_n) \leq \sup \mathcal{R}(f \mid T_d)$ for each such n, and so by Corollary 5.15,

$$(1) \qquad \inf \mathcal{R}(f \mid T_d) \leq \lim_n f(X_n) \leq \sup \mathcal{R}(f \mid T_d).$$

Applying Theorem 7.20 to (1) we see that

$$\liminf_{x \to a} f(x) \leq \lim_n f(X_n) \leq \limsup_{x \to a} f(x).$$

7.74   Theorem.   If f satisfies the conditions of the preceding theorem, then there exists a sequence X such that $\mathcal{R}(X) \subset (S - \langle a \rangle)$, $\lim_n X_n = a$, and $\lim_n f(X_n) = \limsup_{x \to a} f(x)$ $\left(\lim_n f(X_n) = \liminf_{x \to a} f(x)\right)$.

Proof.   We let $A = \limsup_{x \to a} f(x)$ $\left(A = \liminf_{x \to a} f(x)\right)$. In the notation of § 7.72, which we shall use throughout this theorem, we see that

$$(1) \qquad A = \lim_{d \to 0+} G(d) = \lim_{d \to 0} G(d) \left(A = \lim_{d \to 0+} g(d) = \lim_{d \to 0} g(d)\right).$$

---

[14] Since g and G are defined for positive numbers only, we may write $d \to 0$ instead of $d \to 0+$ in the limit expressions involving g and G.

We consider the sequence $\{G(1/n)\}$ $(\{g(1/n)\})$ and use Lemma 7.4 and (1) to conclude that

$$(2) \qquad \lim_n G\left(\frac{1}{n}\right) = A, \qquad \left(\lim_n g\left(\frac{1}{n}\right) = A\right).$$

We define the sequence V (W) so that for each $n \in P$,

$$V_n = \left\{x \mid x \in T_{1/n} \text{ and } G\left(\frac{1}{n}\right) - \left(\frac{1}{n}\right) < f(x) \le G\left(\frac{1}{n}\right)\right\} \text{ if } G\left(\frac{1}{n}\right) \text{ is finite};$$

$$V_n = \{x \mid x \in T_{1/n} \text{ and } n < f(x) \le +\infty\} \text{ if } G\left(\frac{1}{n}\right) = +\infty;$$

$$V_n = \{x \mid x \in T_{1/n} \text{ and } -\infty \le f(x) < -n\} \text{ if } G\left(\frac{1}{n}\right) = -\infty.$$

$$(3)$$

$$\left(W_n = \left\{x \mid x \in T_n \text{ and } g\left(\frac{1}{n}\right) \le f(x) < g\left(\frac{1}{n}\right) + \frac{1}{n}\right\} \text{ if } g\left(\frac{1}{n}\right) \text{ is finite};$$

$$W_n = \{x \mid x \in T_n \text{ and } n < f(x) \le +\infty\} \text{ if } g\left(\frac{1}{n}\right) = +\infty;$$

$$W_n = \{x \mid x \in T_n \text{ and } -\infty \le f(x) < -n\} \text{ if } g\left(\frac{1}{n}\right) = -\infty.\right)$$

It follows from the definition of $G(1/n)$ $(g(1/n))$ and Theorem 5.5 that for each positive integer n there exists $y \in \mathcal{R}(f \mid T_{1/n})$ and therefore $x \in T_{1/n}$ such that $y = f(x)$ and $f(x)$ satisfies the appropriate one of the expressions in (3) that defines $V_n(W_n)$, depending on whether $G(1/n)$ is finite, $G(1/n) = +\infty$, or $G(1/n) = -\infty$ $(g(1/n)$ is finite, $g(1/n) = +\infty$, or $g(1/n) = -\infty)$. Hence $x \in V_n (x \in W_n)$ and $V_n \ne \varnothing$ $(W_n \ne \varnothing)$ for each $n \in P$. From the axiom of choice as expressed in § 1.51, we infer the existence of a sequence X such that $X_n \in V_n (X_n \in W_n)$ for each $n \in P$. Clearly $\mathcal{R}(X) \subset (S - \langle a \rangle)$ and $\lim_n X_n = a$ due to the definition of V (W).

If $A \in R$, then, as we saw in the proof of Theorem 5.11, there exists a positive integer N such that $G(1/n)$ $(g(1/n))$ is a finite real number whenever $N < n \in P$. From our definition of V_n (W_n) it follows that

$$G\left(\frac{1}{n}\right) - \frac{1}{n} < f(X_n) \le G\left(\frac{1}{n}\right) \qquad \left(g\left(\frac{1}{n}\right) \le f(X_n) < g\left(\frac{1}{n}\right) + \frac{1}{n}\right)$$

for each such n. Applying Theorem 5.13 and (2) to this last relation we conclude that $\lim_n f(X_n) = A$.

If $A = +\infty$, then given any positive number M there exists a positive integer N such that

(4) $$G\left(\frac{1}{n}\right) > M + 1, \qquad \left(g\left(\frac{1}{n}\right) > M + 1\right)$$

whenever $N < n \in P$. Now there exists a positive integer $K \geq M + N$. From our definition of V_n (W_n) we see that if $N < n \in P$, then either

(5) $$G\left(\frac{1}{n}\right) - \frac{1}{n} < f(X_n), \qquad \left(g\left(\frac{1}{n}\right) \leq f(X_n)\right)$$

if $G(1/n)$ $(g(1/n))$ is finite or

(6) $$n < f(X_n)$$

if $G(1/n) = +\infty$ $(g(1/n) = +\infty)$. The possibility that $G(1/n) = -\infty$ $(g(1/n) = -\infty)$ is ruled out by (4).

Thus if $K < n \in P$, we see from (4), (5), and (6) that $f(X_n) > G(1/n) - (1/n) > M$ $(f(X_n) \geq g(1/n) > M)$ or else $f(X_n) > n > M$. In either case $f(X_n) > M$ for each such n, and so $\lim_n f(X_n) = +\infty = A$.

In case $A = -\infty$, it is easily seen by making rather obvious modifications in the proof just given that $\lim_n f(X_n) = -\infty = A$. This completes our proof.

7.75 Theorem. If f satisfies the conditions of Theorem 7.73, then $\lim\limits_{x \to a} f(x) \in Re$ if and only if $\limsup\limits_{x \to a} f(x) = \liminf\limits_{x \to a} f(x) \in Re$; and $\lim\limits_{x \to a} f(x)$ coincides with their common value.

Proof. Suppose $\limsup\limits_{x \to a} f(x) = \liminf\limits_{x \to a} f(x) \in Re$. For any $d > 0$, $g(d) = \inf \mathscr{R}(f \mid T_d)$ and $G(d) = \sup \mathscr{R}(f \mid T_d)$, in the notation of § 7.72, so that if $x \in T_d$,

(1) $$g(d) \leq f(x) \leq G(d).$$

If $A \in R$ and $\epsilon > 0$, there exist numbers $\delta' > 0$ and $\delta'' > 0$ such that

(2) $$|g(d) - A| < \epsilon$$

whenever $0 < d < \delta'$ and

(3) $$|G(d) - A| < \epsilon$$

whenever $0 < d < \delta''$. We let d be a positive number less than both δ' and δ''. If $x \in T_d$ then from (1), (2), and (3) we have

$$A - \epsilon < g(d) \leq f(x) \leq G(d) < A + \epsilon,$$

whence $|f(x) - A| < \epsilon$. Since $N_d(a) \cup \langle a \rangle = I$ is a neighborhood of a and $T_d = I \cap \mathscr{D}(f) - \langle a \rangle$, it follows that $\lim\limits_{x \to a} f(x) = A$.

If $A = +\infty$ $(-\infty)$, then $\lim_{d \to 0+} g(d) = +\infty$ $\left(\lim_{d \to 0+} G(d) = -\infty\right)$, and if $0 < M < +\infty$ there exists a positive number δ''' such that $g(d) > M$ $(G(d) < -M)$ whenever $0 < d < \delta'''$. We let d be such a number; then if $x \in T_d = I \cap \mathscr{D}(f) - \langle a \rangle$ it follows from (1) and these last relations that $f(x) \geq g(d) > M$ $(f(x) \leq G(d) < -M)$ and so $\lim_{x \to a} f(x) = +\infty = A$ $\left(\lim_{x \to a} f(x) = -\infty = A\right)$.

To prove the converse we assume that $\lim_{x \to a} f(x) \in Re$. Recall that this requires $\lim_{n} f(Y_n) = \lim_{x \to a} f(x)$ whenever Y is a sequence with $\mathscr{R}(Y) \subset (S - \langle a \rangle)$ and $\lim_{n} Y_n = a$. Observe that this relation must be true in particular for sequences X' and X'' whose existence is guaranteed by Theorem 7.74 such that $\lim_{n} f(X_n') = \liminf_{x \to a} f(x)$, $\lim_{n} f(X_n'') = \limsup_{x \to a} f(x)$, and so conclude that $\liminf_{x \to a} f(x) = \lim_{x \to a} f(x) = \limsup_{x \to a} f(x)$ as required.

Exercises

§§ 7.1 to 7.7

1. Give the separate proofs of Lemma 7.4 in each of the four cases listed under Definition 7.1, that is, take account of the various kinds of neighborhoods that may occur.
2. Given $f(x) = x^2 - 5x$ for all $x \in R$, prove that $\lim_{x \to 2} f(x) = -6$. Determine a value of $\delta > 0$ associated with a given $\epsilon > 0$ in accordance with Definition 7.1.
3. Given $f(x) = x^5 + 3x^4$ for all real x, prove that $\lim_{x \to -1} f(x) = 2$. Find a value of $\delta > 0$ associated with $\epsilon > 0$.
4. Given $a \in R$ and $f(x) = x^2$ for all $x \in R$, prove that $\lim_{x \to a} f(x) = a^2$. Find a value of $\delta > 0$ associated with $\epsilon > 0$.
5. Given $f(x) = 1/x$ for all real $x \neq 0$, show that $\lim_{x \to 1} f(x) = 1$. Find $\delta > 0$ associated with $\epsilon > 0$. (*Hint:* Note that if $x > \frac{1}{2}$, then $|1/x| < 2$. Then take an initial restriction $|x - 1| < \frac{1}{2}$ to exploit this fact, and proceed from this point as in the worked examples.)[15]
6. Given $f(x) = x/(x + 1)$ for all real $x \neq -1$, prove that $\lim_{x \to 0} f(x) = 0$. Find $\delta > 0$ associated with $\epsilon > 0$.
7. Given $f(x) = x/(x^2 + 1)$ for all real x, prove that $\lim_{x \to -2} f(x) = -\frac{2}{5}$. Find $\delta > 0$ associated with $\epsilon > 0$.
8. Given f, so defined that $a \in R$ is a cluster point of $\mathscr{D}(f)$, $A \in R$, $k \in R$, $\eta > 0$, and $|f(x) - A| \leq |k| \cdot |x - a|$ whenever $0 < |x - a| < \eta$ and $x \in \mathscr{D}(f)$. Prove that $\lim_{x \to a} f(x) = A$.

---

[15] It is necessary to stay at some positive fixed distance from 0 in order to control the size of $1/x$. This could not be done with an initial restriction $|x - 1| < 1$, since then $0 < x < 1$, and x could be arbitrarily close to zero.

9. Given $f(x) = x \sin (1/x)$ for all real $x \neq 0$, prove that $\lim_{x \to 0} f(x) = 0$. Find $\delta > 0$ associated with $\epsilon > 0$. You may assume known any elementary properties of the sine function you may require.

10. Given $f(x) = (x^2 - 4)/(x + 2)$ for all real $x \neq -2$, show that $\lim_{x \to -2} f(x) = -4$. Find $\delta > 0$ associated with $\epsilon > 0$.

11. Given $f(x) = (x^3 + 3x^2 - 4x - 12)/(x - 2)$ for all real $x \neq 2$, show that $\lim_{x \to 2} f(x) = 20$.

12. Given f so defined on R that $f(x) = 0$ if x is rational, $f(x) = 1$ if x is irrational, show that f has no limit at 0. (*Hint:* Find two sequences X and X' converging to zero, but never taking the value zero, with $\lim_{n} f(X_n) = 0$, $\lim_{n} f(X'_n) = 1$.)

13. Given that $f(x) = \sin (1/x)$ for all real $x \neq 0$, show that f has no limit at 0.

14. Given f so defined on R that

$$f(x) = x^3 + 3 \qquad \text{if } x \leq 0$$
$$f(x) = 3(x + 1) \qquad \text{if } x > 0,$$

prove that $\lim_{x \to 0} f(x) = 3$. (*Hint:* Find different values of $\delta > 0$ associated with $\epsilon > 0$, one for positive and the other for negative values of x. Take the smaller as your choice for the required δ.)

15. Suppose that A and A' are finite limits of f at a. Show directly from the definition of functional limits that $A = A'$. (*Hint:* The proof can be modeled on that given in Lemma 4.3. Note that for any $x \in \mathscr{D}(f)$ for which $f(x) \in R$ $|A - A'| \leq |A - f(x)| + |A' - f(x)|$ and that each of the terms on the right can be made arbitrarily small by suitably restricting x.

16. Show directly from the definition of functional limits that if A and A' are limits of f at a, then it is not possible for A to be finite and A' to be infinite. (*Hint:* Show that if A is finite, then f must be bounded in some neighborhood of a, i.e., there exists $\delta > 0$ and a finite number $M \geq 0$ such that $|f(x)| \leq M$ for all x satisfying $0 < |x - a| < \delta$, $x \in \mathscr{D}(f)$. Show also that if A' is infinite, then $|f(x)| > M$ for all x satisfying $0 < |x - a| < \delta'$, $x \in \mathscr{D}(f)$, for some $\delta' > 0$. Derive a contradiction.)

17. Show directly from the definition of functional limits that if A and A' are infinite limits of f at a, then it cannot happen that $A = +\infty$ and $A' = -\infty$. (*Hint:* Find numbers $\delta > 0$ and $\delta' > 0$ such that $f(x) > 0$ for all x satisfying $0 < |x - a| < \delta$, $x \in \mathscr{D}(f)$, and $f(x) < 0$ for all x satisfying $0 < |x - a| < \delta'$, $x \in \mathscr{D}(f)$. Obtain a contradiction from these facts. Putting the result of this problem together with those in Problems 15 and 16 gives us a proof of the uniqueness of functional limits.)

§ 7.8

18. Given $f(x) = x/(x + 3)^2$ for all real $x \neq -3$, prove that $\lim_{x \to -3} f(x) = -\infty$. Find a value of $\delta > 0$ corresponding to an arbitrary finite $M > 0$ in accordance with Definition 7.1, Case 2.

19. Given $f(x) = (x^3 + 1)/(x + 1)^3$ for each real $x \neq -1$, show that $\lim_{x \to -1} f(x) = +\infty$. Find $\delta > 0$ corresponding to $M > 0$.

20. Given $f(x) = 1/(x - 1)$ for all real $x > 1$, that is, $\mathscr{D}(f) = \{x \mid 1 < x < +\infty\}$, show that $\lim_{x \to 1} f(x) = +\infty$. Find $\delta > 0$ corresponding to $M > 0$.

21. Given $g(x) = 1/(x - 1)$ for all real $x < 1$, show that $\lim_{x \to 1} g(x) = -\infty$. Find $\delta > 0$ corresponding to $M > 0$.

22. Let $h(x) = 1/(x - 1)$ for all real $x \neq 1$. Show that h has no limit at 1. (*Hint:* Consider a sequence X consisting solely of numbers less than 1, $\lim_n X_n = 1$, and a sequence X' consisting solely of numbers greater than 1, $\lim_n X_n' = 1$. Consider the sequences $\{h(X_n)\}$ and $\{h(X_n')\}$ in the light of Problems 20 and 21.)

23. Given $f(x) = (10/x) - (1/x^2)$ for all real $x \neq 0$, show that $\lim_{x \to 0} f(x) = -\infty$. Find $\delta > 0$ corresponding to $M > 0$.

24. Let f be so defined that $f(x) = 1/10x^2$ if $x > 0$, $f(x) = -5/x$ if $x < 0$. Show that $\lim_{x \to 0} f(x) = +\infty$. Find $\delta > 0$ corresponding to $M > 0$. (*Hint:* Find two values of δ corresponding, respectively, to positive and negative values of x. Then take the smaller of these for the desired number δ.)

25. Let $f(x) = |1/(1 - x^2)^2|$ for all real $x \neq 1$ and -1. Show that $\lim_{x \to 1} f(x) = \lim_{x \to -1} f(x) = +\infty$. Find $\delta > 0$ corresponding to $M > 0$.

26. Let f be so defined that $f(x) = 1/x^3$ if $0 \neq x$ is rational and $f(x) = 1/x^2$ if x is irrational. Show that f has no limit at 0.

27. Let f be defined on a set $S \subset Re$ of which $a \in R$ is a cluster point. Suppose there exists finite positive numbers k and η such that $f(x) > k/|x - a|$ for all $x \in \mathscr{D}(f)$ satisfying $0 < |x - a| < \eta$. Show that $\lim_{x \to a} f(x) = +\infty$.

28. Show by constructing suitable examples that if $a \in R$ and $\lim_{x \to a} |f(x)| = +\infty$, it is not necessarily true that either $\lim_{x \to a} f(x) = +\infty$ or $\lim_{x \to a} f(x) = -\infty$.

§ 7.9

29. Let $f(x) = 1/(1 + x^2)$ for all real x. Prove that $\lim_{x \to +\infty} f(x) = \lim_{x \to -\infty} f(x) = 0$. Find $K > 0$ associated with $\epsilon > 0$ in accordance with Definition 7.1.

30. Given $f(x) = x/(x + 2)^2$ for all real $x \neq -2$. Prove that $\lim_{x \to +\infty} f(x) = \lim_{x \to -\infty} f(x) = 0$. Find $K > 0$ associated with $\epsilon > 0$.

31. Given $f(x) = \dfrac{x + 1}{2x^2 - 1}$ for all real $x \neq \pm\sqrt{2}/2$. Prove that $\lim_{x \to +\infty} f(x) = \lim_{x \to -\infty} f(x) = 0$. Find $K > 0$ associated with $\epsilon > 0$.

32. Given $f(x) = \sin x$ for all real x, show that f has no limit at either $+\infty$ or $-\infty$. (*Hint:* Find sequences X and X' of finite numbers with limit $+\infty$ and $-\infty$, respectively, such that $\lim_n f(X_n)$ and $\lim_n f(X_n')$ exist but are unequal. You may assume necessary elementary properties of the sine function.)

33. Let $f(x) = \dfrac{2x^3 - x^2}{x^3 + 2x^2 + 1}$. Prove that $\lim_{x \to +\infty} f(x) = \lim_{x \to -\infty} f(x) = 2$. Find $K > 0$ associated with $\epsilon > 0$.

34. Let $f(x) = 1/3x$, for all rational $x \neq 0$, and $f(x) = 1/x^2$ for all irrational x. Prove that $\lim_{x \to +\infty} f(x) = \lim_{x \to -\infty} f(x) = 0$. Find $K > 0$ associated with $\epsilon > 0$.

35. Given $f(x) = 1$ for all rational x, $f(x) = 0$ for all irrational x, prove that f has no limit at $+\infty$ nor at $-\infty$.

36. Let $n \in P$ and $f(x) = 1/x^{1/n}$ for each real $x > 0$. Recall Problems 62 and

63 of the Exercises at the end of Chapter 2, and show that $\lim\limits_{x\to+\infty} f(x) = 0$. Find $K > 0$ associated with $\epsilon > 0$.

37. Given $f(x) = 1/\sqrt{x^2 - 4}$ for all real x such that $x^2 > 4$, prove that $\lim\limits_{x\to+\infty} f(x) = \lim\limits_{x\to-\infty} f(x) = 0$. Find $K > 0$ associated with $\epsilon > 0$.

38. Suppose that $a \in R$. Show that $\lim\limits_{x\to+\infty} 1/(x + a) = \lim\limits_{x\to-\infty} 1/(x + a) = 0$. Let $n \in P$. Show that for all x satisfying either $x > |a| + 1$ or $x < -(|a| + 1)$, the relation $|1/(x + a)^n| < 1/|x + a|$ holds. Then prove that $\lim\limits_{x\to+\infty} 1/(x + a)^n = \lim\limits_{x\to-\infty} 1/(x + a)^n = 0$.

§ 7.10

39. Given $f(x) = x^2 - 20x - 100$ for all real x, show that $\lim\limits_{x\to+\infty} f(x) = \lim\limits_{x\to-\infty} f(x) = +\infty$. Determine a value of $K > 0$ associated with $M > 0$ by Definition 7.1 in each case.

40. Given $f(x) = x^2/(x + 10)$ for all real $x \neq -10$, show that $\lim\limits_{x\to+\infty} f(x) = +\infty$, $\lim\limits_{x\to-\infty} f(x) = -\infty$. Find a value of $K > 0$ associated with $M > 0$ in each case.

41. Let $f(x) = (x^2 - 10x - 1)/(x + 5)$ for all real $x \neq -5$. Show that $\lim\limits_{x\to+\infty} f(x) = +\infty$, $\lim\limits_{x\to-\infty} f(x) = -\infty$. Find a value of $K > 0$ associated with $M > 0$ in each case.

42. Given $f(x) = \sqrt{x}$ for all real $x \geq 0$. Prove that $\lim\limits_{x\to+\infty} f(x) = +\infty$. Find $K > 0$ associated with $M > 0$. (*Hint:* Recall Problem 36 above.)

43. Given $f(x) = \sqrt{x^2 - 4}$ for all real x such that $|x| \geq 2$, show that $\lim\limits_{x\to+\infty} f(x) = \lim\limits_{x\to-\infty} f(x) = +\infty$. Find $K > 0$ associated with $M > 0$.

44. Given $f(x) = x^2$ for all non-integral real values of x, and $f(x) = 1000x$ for all integral values of x, prove that $\lim\limits_{x\to+\infty} f(x) = +\infty$. Find $K > 0$ associated with $M > 0$. Also prove that f does not have a limit at $-\infty$.

45. Let $f(x) = 2^x$ for all positive integral values of x and $f(x) = 1000x$ for all other real values of x. Prove that $\lim\limits_{x\to+\infty} f(x) = +\infty$. Find $K > 0$ associated with $M > 0$.

46. Let $f(x) = x$ if x is rational and $f(x) = -x$ if x is irrational. Prove that f has no limit at $+\infty$ nor at $-\infty$.

47. Let $f(x) = x$ if x is rational and $f(x) = -|x|$ if x is irrational. Show that $\lim\limits_{x\to-\infty} f(x) = -\infty$, but that f has no limit at $+\infty$.

48. Let $f(x) = +\infty$ if x is non-integral and $f(x) = |x|^x$ if x is an integer. Show that $\lim\limits_{x\to+\infty} f(x) = +\infty$, but that f has no limit at $-\infty$. You may assume that the properties of negative exponents are known.

49. Let $f(x) = x \sin x$ for all real x. Show that f has no limit at $+\infty$ nor at $-\infty$. You may assume known elementary properties of the sine function.

§§ 7.11 to 7.30

50. Prove Theorem 7.25 directly from the definition of limits of functions of a real variable without using theorems on sequential limits.

51. Prove Theorem 7.26 (i) and (ii) directly from Definition 7.1. (*Hint:* Look at the technique used in Theorem 4.16 and adapt it to the present situation.)

52. Prove that if f is a function, $\mathcal{D}(f) \subset Re$, $\mathcal{R}(f) \subset Re$, a is a cluster point of $\mathcal{D}(f)$, I is a neighborhood of a and $f(x) = x$ whenever $a \neq x \in I \cap \mathcal{D}(f)$, then $\lim_{x \to a} f(x) = a$. Under the same hypotheses, show that $\lim_{x \to a} (f(x))^n = a^n$ for each positive integer n.

53. Use mathematical induction to extend (i) and (iii) of Theorem 7.26 to sums and products of n functions.

54. Assuming that all the conditions of Problem 52 are satisfied and $f(x) = b_0 x^n + b_1 x^{n-1} + \ldots + b_n$ for all $x \in R$ and the coefficients are all real, show that $\lim_{x \to a} f(x) = b_0 a^n + b_1 a^{n-1} + \ldots + b_n$ using the results of Problem 52 and 53.

55. Show that if $f(x) = a_0 + a_1/x + a_2/x^2 + \cdots + a_n/x^n$ for all $x \in R$, then $\lim_{x \to +\infty} f(x) = \lim_{x \to -\infty} f(x) = a_0$.

56. Let f be so defined on R that for each $x \in R$ except for the roots of the denominator,

$$f(x) = \frac{a_0 x^n + a_1 x^{n-1} + \cdots + a_n}{b_0 x^m + b_1 x^{m-1} + \cdots + b_m},$$

where the coefficients are all real, $a_0 \neq 0$, $b_0 \neq 0$.

Show that if $m = n$, then $\lim_{x \to +\infty} f(x) = \lim_{x \to -\infty} f(x) = a_0/b_0$; if $m > n$, then $\lim_{x \to +\infty} f(x) = \lim_{x \to -\infty} f(x) = 0$; if $m < n$, then $\lim_{x \to +\infty} f(x)$ and $\lim_{x \to -\infty} f(x)$ are infinite; their exact signs depend upon the sign of a_0/b_0, and whether $n - m$ is even or odd. (*Hint:* Divide out x^n from the numerator and x^m from the denominator. Then for all large positive and negative values of x,

$$f(x) = x^{n-m} \left(\frac{a_0 + a_1/x + \cdots + a_n/x^n}{b_0 + b_1/x + \cdots + b_m/x^m} \right)$$

Prove that the expression in parentheses tends to a_0/b_0 as x grows large, using Problem 55 and appropriate theorems on limits, and proceed from this point.)

In the following problems, assume that $\lim_{x \to 0} \sin x = 0$, $\lim_{x \to 0} \cos x = 1$, $\lim_{x \to 0} \dfrac{\sin x}{x} = 1$, and the usual identities relating the trigonometric functions. Use any theorems proved thus far to evaluate the limits.

57. Show that $\lim_{x \to 0} \dfrac{\sin x}{\sqrt{x}} = 0$.

58. Show that $\lim_{x \to 0} \dfrac{1 - \cos x}{x} = 0$.

59. Prove that $\lim_{x \to 0} \dfrac{1 - \cos x}{x^2} = \dfrac{1}{2}$.

60. Prove that $\lim_{x \to 0+} \cot x = +\infty$, $\lim_{x \to 0-} \cot x = -\infty$ (Write $\cot x = \cos x/\sin x$).

61. Show that $\lim_{x \to 0} x \cot x = 1$.

§§ 7.31 to 7.36

To evaluate the following limits, you may use any of the theorems established thus far, together with the properties of sines and cosines mentioned above. You may also assume that

$$\lim_{x \to +\infty} a^x = +\infty \text{ if } 1 < a, \text{ and } \lim_{x \to 0} \frac{e^x - 1}{x} = 1.$$

62. Evaluate $\lim\limits_{x \to +\infty} \dfrac{\sin 2x}{x}$; $\lim\limits_{x \to -\infty} \dfrac{\sin 3x}{x}$.

63. Evaluate $\lim\limits_{x \to +\infty} x(1 - \cos(1/x))$.

64. Evaluate $\lim\limits_{x \to +\infty} (xe^{1/x} - x)$.

65. Evaluate $\lim\limits_{x \to -\infty} a^x$ if $0 < a < 1$.

66. Evaluate $\lim\limits_{x \to -\infty} \dfrac{\sin(1/x)}{\sin(2/x)}$.

67. Evaluate $\lim\limits_{x \to 0} \dfrac{\sin(x^2)}{x \sin 5x}$

68. Evaluate $\lim\limits_{x \to 0} \dfrac{\sin^3 2x - \sin 2x}{x \cos^2 x}$.

69. Under the hypotheses of Theorem 7.35 show that if $\lim\limits_{t \to 0+} h(t)$ $(\lim\limits_{t \to 0-} h(t))$ fails to exist, then $\lim\limits_{x \to +\infty} g(x)$ $(\lim\limits_{x \to -\infty} g(x))$ fails to exist.

70. Suppose that f is a function with domain and range contained in Re, $0 < M < +\infty$, $a \in R$ and $|f(x') - f(x'')| < M |x' - x''|$ whenever $0 < |x' - a| < 1$ and $0 < |x'' - a| < 1$. Prove that $\lim\limits_{a} f$ exists and is a finite real number. Can you infer what it is? Give reasons, or else a counterexample to prove your point.

71. If, in Problem 70, f satisfies the modified condition $|f(x') - f(x'')| < M |x'^2 - x''^2|$ whenever $0 < |x' - a| < 1$ and $0 < |x'' - a| < 1$, show that $\lim\limits_{a} f$ exists in R.

§§ 7.37 to 7.47

72. Prove that if f is a polynomial defined on a set $S \subset R$, then f is continuous at x whenever $x \in S$. (*Hint:* Recall Problem 54 above.)

73. Prove that if f and g are polynomials with a common domain $S \subset R$, $a \in S$, and $g(a) \neq 0$, then f/g is continuous at a.

74. Given that $f(x) = x \sin 1/x$ for each real number $x \neq 0$, prove that there exists a way of defining f at 0 so that the extended function is continuous at 0. Assume any properties of the sine function you may require.

75. If $f(x) = x^2$ for all real $x \leq 2$ and $f(x) = 12 - x^3$ for all real $x > 2$, show that f is continuous at $x = 2$. Prove that f is continuous for all finite real values of x.

76. Given that $f(x) = 0$ for all rational values of x, $f(x) = 1$ for all irrational values of x, show that f is discontinuous (i.e., not continuous) for all real values of x.

77. If $f(x) = \sin 1/x$ for all real $x = 0$, show that it is impossible to define f at 0 so that the extended function is continuous at 0.

78. Let f be so defined that $f(x) = x$ for each rational number x, $f(x) = 0$ for

each irrational number x. Show that f is continuous at 0 but not anywhere else.

79. We let f be a function continuous on its entire domain $I = \{x \mid a < x < b\}$, where a and b are finite real numbers and $a < b$, and we assume that $\lim_{a} f$ is a finite real number. Show that f may be continued onto the interval $J = \{x \mid a \le x < b\}$ so as to be continuous on all of J, that is, there exists a function g agreeing with f on I and continuous on J.

80. We suppose that f is a function whose domain S and range are subsets of Re, a is a finite member of S, and that S is the union of two sets S' and S'' to both of which a belongs. Show that if $(f \mid S')$ and $(f \mid S'')$ are both continuous at a, then so is f. (This is a partial converse of Theorem 7.42).

81. Let f be a function whose domain T is a subset of R. Suppose that S is a finite sequence (cf. opening remarks of Chapter 4) of subsets of R, whose domain is $\hat{n}$, where $n \in P$. Assume that $a \in S_i$ and that $(f \mid S_i)$ is continuous at a for each $i \in \hat{n}$, and that T is the union of the range of S. Show by induction on the result of Problem 80 that f is necessarily continuous at a.

82. If S is an infinite sequence of subsets of R, but all other assumptions of Problem 81 are valid, either prove that f must be continuous at a or else show by a suitable counterexample that f need not be continuous at a.

83. If $b \in R$, $\varnothing \ne S \subset R$, and f is so defined that $f(x) = b$ whenever $x \in S$, then f is continuous on S.

§§ 7.48 to 7.75

84. Given $f(x) = x^3 - 2x + 3$ for each $x \in I$, where $I = \{x \mid -3 \le x \le 1\}$, show that f is uniformly continuous on I. Actually determine a value of δ associated with an arbitrary $\epsilon > 0$.

85. Given $f(x) = \dfrac{2x^2 - 1}{x + 2}$ for each $x \in I$, where $I = \{x \mid -1 \le x \le 5\}$, show that f is uniformly continuous on I by finding a value of δ associated with an arbitrary $\epsilon > 0$.

86. Given $f(x) = 1/x$ for all $x \in I$, where $I = \{x \mid 0 < x \le 2\}$, show that f is not uniformly continuous on I.

87. Let f be a function uniformly continuous on its domain S, of which a is a cluster point, but $a \notin S$. Show that there exists a continuation g of f that is defined and continuous on $S \cup \langle a \rangle$, that is, g coincides with f on S and is continuous on $S \cup \langle a \rangle$.

88. Let u be the sequence of functions defined on $S = \{x \mid -2 \le x \le 1\}$, such that $u_n(x) = \dfrac{nx^3}{n + 1}$ for each $n \in P$ and each $x \in S$. Prove that u converges uniformly on S to the function U defined so that $U(x) = x^3$ for each $x \in S$.

89. Let u be the sequence of functions defined on $S = \{x \mid 0 < x < +\infty\}$ by $u_n(x) = (1/n) \sin 1/x$ for each $n \in P$ and each $x \in S$. Show that the sequence u converges uniformly to 0 on S.

90. Let u be the sequence of functions defined on R by $u_n(x) = 0$ if x is irrational, $u_n(x) = 1/n$ if x is rational, for each $n \in P$. Prove that u converges uniformly to 0 on R.

91. Given the sequence u on $I = \{x \mid 0 \le x \le 1\}$ defined by $u_n(x) = x^n$ for each $n \in P$ and each $x \in I$, prove that the sequence u converges to U, where $U(x) = 0$ if $0 \le x < 1$ and $U(1) = 1$. Using Theorem 7.69, show that the convergence of the sequence u is not uniform on I.

92. Let f be a function defined on $S \subset Re$, taking values in Re. Assume that a is a cluster point of S. Let $k \in Re$ and suppose that there is a neighborhood I of a, such that $k \le f(x)$ $(f(x) \le k)$ for each x, $a \ne x \in I \cap S$. Show that $k \le \liminf_{x \to a} f(x)$ $(\limsup_{x \to a} f(x) \le k)$.

93. In Exercise 91, let $U(x) = \lim_n u_n(x)$ for each $x \in I$. Show that for each $a \in I$, except $a = 1$, it is true that $\lim_{x \to a}(\lim_n u_n(x)) = U(a)$ and $\lim_n(\lim_{x \to a} u_n(x)) = U(a)$, and so the iterated limits are equal. Show that this is not true at $a = 1$.

94. For each positive integer n, we define the function f_n on the closed interval I with endpoints at 0 and 1 by $f_n(x) = x^n/n$ for each $x \in I$. Prove that the sequence $\{f_n\}$ converges uniformly on I to a function identically zero on I. (*Hint:* Show that $\lim_n f_n(1) = 0$, and hence given any $\epsilon > 0$ there exists $N \in P$ such that $|f_n(1)| < \epsilon$ whenever $N < n \in P$. Show that the positive integer N thus associated with ϵ and the point $x = 1$ is associated with ϵ for any point $x \in I$.)

95. For each positive integer n, we define the function f_n on the closed interval I with endpoints at 0 and 1 as follows: if $0 \le x \le 1/2n$, then the value of $f_n(x)$ is on the line segment joining $(0, 0)$ to $(1/2n, 1)$; if $1/2n \le x \le 1/n$, then the value of $f_n(x)$ is on the line segment joining $(1/2n, 1)$ to $(1/n, 0)$; if $1/n \le x \le 1$, then $f_n(x) = 0$. Show that the sequence $\{f_n\}$ converges on I to a function that is identically zero on I. Show also that although the functions of the sequence $\{f_n\}$ are all continuous on I and their limit function is continuous on I, the convergence is not uniform on I.

Suggested Reading List

1. *Abstract Set Theory*, A. A. Fraenkel, North Holland Publishing Company, Amsterdam, 1953.
2. *Introduction to Logic*, A. Tarski, Oxford University Press, New York, 1961.
3. *Sets, Sequences and Mappings*, K. W. Anderson and Dick Wick Hall, John Wiley and Sons, New York, 1963.
4. *Treatise on Advanced Calculus*, P. Franklin, John Wiley and Sons, New York, 1940.
5. *Advanced Calculus*, W. Fulks, John Wiley and Sons, New York, 1960.
6. *Intermediate Analysis*, J. M. H. Olmsted, Appleton-Century Crofts, New York, 1956.

Answers to Problems

1. 1, 2, 3.

2. 1, 2, 3, 4, 5, 6, 7, 8, 9, 10, 11.

3. All even positive integers and no other objects are its members.

4. 1, 2, 3, 4.

5. -1 is a member; 1 and $1\frac{1}{2}$ are not.

6. $\{x \mid x = 0\}$.

7. $\{x \mid x = -1 \text{ or } x = 1\}$.

8. $\{t \mid t = x \text{ or } t = y \text{ or } t = z\}$.

9. $\{x \mid x \text{ is a positive integer and } x \le 50\}$.

10. $\{x \mid \text{there exists a positive integer } n \text{ such that } x = n^2\}$.

11. $\{x \mid 5x^6 - 2x^4 + x + 10 = 0\}$.

12. $\{x \mid \text{there exists a positive integer } n \text{ such that } x = -1/n\}$.

13. $\{x \mid \text{there exists a positive integer } n \text{ such that } x = 2^n\}$.

14. $\{x \mid \text{there exist positive integers } m \text{ and } n \text{ such that } x = m^n\}$.

15. It has no members.

19. t is not a member; if $w < t$, then w is not a member.

20. s and t are members of the set.

22. $x^2 - 4x + 1 = 0$.

§1.1 to 1.5

28. Not true for all A, B, C.

29. Not true for all A, B, C.

30. Not true for all A, B, C.

33. $B \subset A$ and $A \not\subset B$. The members of A are s and t; the only member of B is s.

34. (a) A and B are the only members of C.
(b) The statements "$A \subset C$" and "$B \subset C$" are both false.
(c) "$1 \in C$" and "$2 \in C$" are false statements.

36. $s \in A$ and $t \in A$. The set of objects not belonging to A is $\varnothing$, the empty set.

§1.6 to 1.19

39. Always true.

40. Not true for all sets A, B, C.

41. Not true for all sets A, B, C.

42. Always true.

43. Always true.

44. Always true.

45. Not true for all sets A, B, C.

46. Always true.

47. Always true.

48. Not true for all A, B.

49. Not true for all A, B, C.

50. Not true for all A, B, C.

51. Always true.

52. Always true.

53. Not true for all A, B, C.

54. Not true for all A, B, C.

56. Always true.

57. Always true.

58. Always true.

§1.20 to 1.24

68. $\cup \mathscr{G}$ is the set of all real numbers; $\cap \mathscr{G}$ is the set whose only member is 0.

69. $\mathscr{F}$ may be the null set or the set whose only member is the null set.

70. $\cup \mathscr{F} = A$; $\cap \mathscr{F} = \varnothing$.

§1.25 to 1.36

72. $\mathscr{D}(Q) = \{x \mid -1 \leq x \leq 1\}$;
$\mathscr{R}(Q) = \{y \mid -2 \leq y \leq 2\}$;
Q and Q^{-1} are not functions.

73. $\mathscr{D}(Q) = \{x \mid x \text{ is a non-negative real number}\}$;
$\mathscr{R}(Q) = \{y \mid y \text{ is a real number}\}$;
Q is not a function; $Q^{-1}(x) = x^2$ whenever $x \in \mathscr{R}(Q)$.

74. $\mathscr{D}(Q) = \{x \mid -1 \leq x \leq 2\}$;
$\mathscr{R}(Q) = \{y \mid -4 \leq y \leq 5\}$;
$Q(x) = x^2 - 4x$ whenever $x \in \mathscr{D}(Q)$;
$Q^{-1}(x) = 2 - \sqrt{4 + x}$ whenever $x \in \mathscr{R}(Q)$.

75. $\mathscr{D}(Q) = \{x \mid 0 \leq x \leq 2\}$;
$\mathscr{R}(Q) = \{y \mid 0 \leq y \leq 6\}$;
  $Q(x) = 3 \mid x \mid = 3x$ whenever $x \in \mathscr{D}(Q)$;
  $Q^{-1}(x) = x/3$ whenever $x \in \mathscr{R}(Q)$.

76. Q and Q^{-1} are not functions.

77. $\mathscr{D}(Q) = \{x \mid x$ is a positive integer$\}$;
  $\mathscr{R}(Q) = \{y \mid y$ is a positive integral power of 2$\}$;
  $Q(x) = 2^x$ whenever $x \in \mathscr{D}(Q)$;
  $Q^{-1}(x) = \log_2 x$ whenever $x \in \mathscr{R}(Q)$.

78. $\mathscr{D}(Q) = \{x \mid -1 \leq x \leq 3\}$;
  $\mathscr{R}(Q) = \{y \mid 0 \leq y \leq 6\}$;
  $Q(x) = x^2$ if $-1 \leq x \leq 0$; $Q(x) = 2x$ if $0 \leq x \leq 3$;
  Q^{-1} is not a function.

79. $\mathscr{D}(Q) = \{x \mid x = 5\}$;
  $\mathscr{R}(Q) = \{y \mid y = 2\}$;
  $Q(x) = 2$ whenever $x \in \mathscr{D}(Q)$, that is, if $x = 5$;
  $Q^{-1}(x) = 5$ whenever $x \in \mathscr{R}(Q)$, that is, if $x = 2$.

80. $\mathscr{D}(Q) = \{x \mid x$ is a real number and $x \neq -1\}$;
  $\mathscr{R}(Q) = \{y \mid y$ is a real number and $y \neq 0\}$;
  $Q(x) = 1/(1 + x)$ whenever $x \in \mathscr{D}(Q)$;
  $Q^{-1}(x) = (1 - x)/x$ whenever $x \in \mathscr{R}(Q)$.

81. $\mathscr{D}(Q) = \{x \mid 0 < x < 1\}$;
  $\mathscr{R}(Q) = \{y \mid y$ is a negative real number$\}$;
  $Q(x) = (2/3)x - 2/(3x)$ whenever $x \in \mathscr{D}(Q)$;
  $Q^{-1}(x) = (3x + \sqrt{9x^2 + 16})/4$ whenever $x \in \mathscr{R}(Q)$.

82. $\mathscr{D}(Q) = \mathscr{R}(Q) = \{x \mid x$ is a set$\}$;
  Q and Q^{-1} are not functions.

83. $\mathscr{D}(Q)$ is the set of all circles in the fixed plane;
  $\mathscr{R}(Q) =$ is the set of all points in that plane;
  $Q(x) = $ (center of x) whenever $x \in \mathscr{D}(Q)$;
  Q^{-1} is not a function.

84. $\mathscr{D}(Q) = \{x \mid x$ is a set$\}$;
  $\mathscr{R}(x) = \{y \mid y$ is a singleton whose only member is a set$\}$;
  $Q(x) = \langle x \rangle$ whenever $x \in \mathscr{D}(Q)$;
  $Q^{-1}(x) = $ (the only member of x) $= \cap x$ whenever $x \in \mathscr{R}(Q)$.

85. $\mathscr{D}(Q) = \{x \mid x$ is a set$\} = \mathscr{R}(Q)$;
  Q and Q^{-1} are not functions.

86. $\mathscr{D}(Q) = \{x \mid x$ is a set and $\varnothing \in x\}$;

$\mathscr{R}(Q) = \{y \mid y$ is a set$\}$;

Q is not a function;

$Q^{-1}(x) = x \cap \langle \varnothing \rangle$ whenever $x \in \mathscr{R}(Q)$.

87. $\mathscr{D}(Q) = \{u \mid u$ is a set$\}$;

$\mathscr{R}(Q) = \{y \mid$ there exists a set u such that $y = A \times u\}$;

$Q(x) = A \times x$ whenever $x \in \mathscr{D}(R)$;

$Q^{-1}(x) = $ (the only set y such that $A \times y = x) = \cap \{z \mid A \times z = x\}$
whenever $x \in \mathscr{R}(Q)$.

88. $\mathscr{D}(Q) = \{x \mid x$ is a set$\} = \mathscr{R}(Q)$;

Q and Q^{-1} are not functions.

89. $\mathscr{D}(Q) = \{x \mid x$ is a real number$\}$;

$\mathscr{R}(Q) = \{y \mid y$ is a non-negative real number$\}$;

$Q(x) = x^4$ whenever $x \in \mathscr{D}(Q)$;

Q^{-1} is not a function.

§1.37 to 1.50

101. $g \circ f = \{z \mid z = (1, 0)$ or $z = (5, -2)$ or $z = (6, 1)\}$;

$(g \circ f)^{-1} = \{z \mid z = (0, 1)$ or $z = (-2, 5)$ or $z = (1, 6)\}$.

103. $(f \circ g)(x) = \langle x \cup \langle x \rangle \rangle$;

$(g \circ f)(x) = \langle x \rangle \cup \langle \langle x \rangle \rangle$; $(f \circ g)(x) \neq (g \circ f)(x)$.

104. $(g \circ g)(x) = (x \cup \langle x \rangle) \cup \langle x \cup \langle x \rangle \rangle$.

CHAPTER 2

22. $\{x \mid x < -1$ or $6 < x\}$.

23. $\{x \mid -5 < x < 1\}$.

24. $\{x \mid x < -1$ or $0 < x < 4\}$.

25. $\{x \mid x < -2$ or $1 < x < 5\}$.

26. $\{x \mid x < 1$ or $2 < x\}$.

35. $\{x \mid x < -3\}$.

36. $\{x \mid 1 < x < 3$ or $-3 < x < -1\}$.

37. $\{x \mid x < -1$ or $1 < x\}$.

38. $\{x \mid x < -\frac{1}{3}$ or $1 < x\}$.

CHAPTER 3

1. Not always true.

2. Not always true.

3. Not always true.

4. Always true.

5. Always true.

6. Always true.

CHAPTER 4

42. The sequence is constant, but it is not possible to decide what the constant value is.

43. Same as 42.

44. Same as 42.

45. Same as 42.

46. A may or may not converge. For instance, the sequence $A = (-1)^n/8$ satisfies the condition

$$|A_m - A_n| \le \frac{1}{4} < \frac{mn}{m+n} \text{ for all } m, n \text{ in } P.$$

CHAPTER 5

18. The sequence has no limit.

19. The sequence has no limit.

20. The limit is $+\infty$.

21. If $x > 1$, the limit is $+\infty$; if $x < -1$, there is no limit.

22. The limit is $+\infty$.

23. The sequence has no limit.

24. The limit is $+\infty$.

25. The limit is $+\infty$.

28. $\limsup_n A_n = 1$; $\liminf_n A_n = 0$

29. $\limsup_n A_n = +\infty$; $\liminf_n A_n = 1$

30. $\limsup_n A_n = +\infty$; $\liminf_n A_n = -\infty$.

31. $\limsup_n A_n = +\infty$; $\liminf_n A_n = -\infty$.

32. $\limsup_n A_n = 1$; $\liminf_n A_n = 0$.

INDEX